澜湄职业教育培训中心暨柬埔寨鲁班工坊系列教材

A Series of Textbooks for Lancang-Mekong Vocational Education Training Center and Cambodia Luban Workshop

加工中心操作工培训教程(初级)

Machining Center Operator's Training Course (Elementary)

主　编　张　龙

Chief editor　ZHANG Long

副主编　赵新杰

Deputy editor　ZHAO XinJie

西安电子科技大学出版社

Introduction to the Course

This book gives a systematical and comprehensive introduction to the knowledge related to Computer Numerical Control (CNC) machining processing and programming basics, the principles of CNC operation, and the mechanical structure of CNC machine tool. The book consists of five modules, including technology foundations, programming, operation, preparation and inspection, and maintenance of machining center. The book offers a rich collection of selected cases which combine practicality and fun, focuses on "case-scenario instruction and project-driven learning" and cultivation of application ability.

This series of textbooks can be used by colleges and universities, vocational colleges, open university, and various training schools in their programs like mechanical design and manufacturing, CNC technology, moldule design and manufacturing, and mechatronics.

图书在版编目(CIP)数据

加工中心操作工培训教程: 初级: 英文 / 张龙主编. — 西安: 西安电子科技大学出版社, 2022.12
ISBN 978-7-5606-6487-3

Ⅰ. ①加⋯　Ⅱ. ①张⋯　Ⅲ. ①加工中心—操作—技术培训—教材—英文　Ⅵ. ①TG659

中国版本图书馆 CIP 数据核字(2022)第 128302 号

策划编辑　刘玉芳　　明政珠
责任编辑　刘玉芳
出版发行　西安电子科技大学出版社(西安市太白南路 2 号)
电　　话　(029)88202421　88201467　　邮　　编　　710071
网　　址　www.xduph.com　　　　　　电子邮箱　　xdupfxb001@163.com
经　　销　新华书店
印刷单位　西安日报社印务中心
版　　次　2022 年 12 月第 1 版　　2022 年 12 月第 1 次印刷
开　　本　787 毫米 × 1092 毫米　　1/16　印　张　8.25
字　　数　186 千字
定　　价　36.00 元
ISBN 978-7-5606-6487-3 / TG
XDUP 6789001-1
***** 如有印装问题可调换 *****

General Foreword

Serving the Belt and Road Initiative of China, the Lancang-Mekong Vocational Education Training Center and Cambodia Luban Workshop is a joint project undertaken by Tianjin Sino-German University of Applied Sciences (TSGUAS) for the Ministry of Foreign Affairs, the Ministry of Education and the Tianjin Municipal People's Government. Based in Cambodia, the project is designed to serve five countries in the Lancang-Mekong area and radiate to other ten ASEAN countries. It integrates functions of vocational training, vocational education, scientific research, cultural inheritance and innovation & entrepreneurship, develops both academic and non-academic education, and operates as a market-oriented international vocational training center.

At the initial stage of the project, 18 training rooms including mechanical processing technology, electrical technology and communication technology were built in three training centers for majors of Mechatronics and Communication Technology, with a total construction area of 6814 m^2 and more than 1600 sets of equipment.

The project is to implement a "three-phase" plan. Based on the specialty construction in the first phase, international tourism, logistics engineering, automobile maintenance, building electricity and other specialties are to be set up in the second phase to carry out technical skills training for Chinese & Cambodian enterprises and Cambodian people. Meanwhile, higher vocational education, applied undergraduate education, joint postgraduate education and other academic educations are to be carried out to explore the systematic talents cultivation of "intermediate and high vocational education, undergraduate education, and postgraduate education for a master's degree and doctoral degree".

Since 2017, as many as 95 articles about the project have been published by mainstream media including People's Daily, Guangming Daily, China Education News, Xinhuanet, etc. at home and abroad. After over two months of field study and research, Tianjin Television produced two feature stories named "Khmer Training", each lasting 30 minutes. The two episodes were broadcast on May 6th, 2019 and May 13th, 2019 respectively, featuring on "set sails the vocational education overseas and shine on the Luban Workshop". They give a full coverage of how TSGUAS teachers brought advanced skills to local areas and how friendship flourished along the Belt and Road Initiative(BRI)—a great contribution to the BRI. On July 18th, 2019, the Royal Government of Cambodia conferred the Officer of the SAHAMETREI Medal to the Secretary of the Party Committee of TSGUAS, and the Knight of the SAHAMETREI Medal to the President and Vice President in charge of this project, with the signature of Prime Minister Hun Sen of Cambodia. On July 22, 2019, China Education Association for International Exchange awarded TSGUAS the medal of "Featured Cooperation Project of China-ASEAN Higher Vocational Colleges". In October 2019, the President of National Polytechnic Institute of Cambodia (NPIC) presented 11 teachers with certificates and medals for their outstanding contributions to the Ministry of Labor and Vocational Training of Cambodia. Tianjin Sino-German University of Applied Sciences together with National Polytechnic Institute of Cambodia (NPIC) and their partners with enterprises was approved as the Belt and Road Joint Laboratory (Research Center) — Tianjin Sino-German and Cambodia Intelligent Motion Device and Communication Technology Promotion Center in December, 2020.

The Center has become a training base in Langcang-Mekong area for technical talents training, a talent support base for Chinese enterprises overseas, a demonstration base for international students, or a base for teacher training. The Center is a key educational project of the Ministry of Foreign Affairs to serve the Belt and Road Initiative with foreign participation and entity institutions involved locally. The project aims to serve the social-economic and cultural development of the countries along the BRI, enhancing the well-being of mankind; it will also serve the production output capacity of Chinese enterprises to help national development as well as to enhance the international development of vocational education and the quality of its connotation. The project is a bridge connecting vocational education of Tianjin with the world, which marks a new stage of the city's international exchange and cooperation from a lower-medium to a medium-higher level.

The team of the project have compiled a series of textbooks for training, involving six occupations (electrotechnics, lathe, milling, CNC operation, bench and 4G communication network) from elementary, intermediate to advanced level based on current human resources in Langcang-Mekong countries, China's teaching equipment, and Chinese vocational qualification standards. These 19 textbooks orient competence development and cultivation of the ability work tasks setting, combining theory with practice, and learning with practicing so as to put knowledge and skills into real situations. The textbooks aim to provide skills standards for the six occupations and lay foundations for the upgrading of the technological level of Lancang-Mekong countries.

ZHANG XingHui

Party Secretary of Tianjin Sino-German University of Applied Sciences

March, 2021

A Series of Textbooks for Lancang-Mekong Vocational Education Training Center and Cambodia Luban Workshop Editorial Committee

Preface

This textbook is designed in line with the appraisal standards for the national professional skills, followed with the requirements of National Professional Standards — Elementary Operators at Machining Center, and arranged and designated based on the technology examination and assessment skills. In completing this book, we have integrated the teaching of technical skills — the machining technology of CNC machining tool, programming and parts-making by CNC machining, etc., into practical training to better serve employment market, which fully displays the characteristics of project-based teaching approach — "teaching, learning and doing". By doing these projects step by step, students can both acquire theoretical knowledge and participate in practical training to improve their perceptual skills, thus achieving more with less efforts. With these characteristics, this book is expected to prepare students to reach the elementary level as CNC machining center operators.

The series of machining center operator's training course consists of three books organized in levels of the elementary, the intermediate, and the advanced level. Unlike some traditional textbooks which focus merely on in-depth learning, it is concerned with how to integrate "teaching, learning and doing" in hoping that students can gain both theoretical knowledge and practical skills. During the process of developing the book series, we strive to follow the practical needs and cut down boring and not-so-practical theories. Moreover, the book series is industry-oriented, ability-oriented, and student-centered, which makes it more practical and forward-looking, and more closely integrated with the job market.

In addition, we try to break through the outdated philosophy of education, and demonstrate "case-based teaching" in the series. While explaining a certain theory, actual situation is combined and a number of clear and practical cases are analyzed, with the aim of composing high-quality books. The cases designed in this textbook can not only facilitate the teaching, but also inspire students to think, improve students' practical skills, and reform the training mode of talents.

It is available to scan the two-dimensional code below for corresponding contents of this book in Chinese.

During compiling of this textbook, it is inevitable that there exist some errors and shortcomings due to our limitations. We look forward to suggestions and advices on how to better revise and improve this book.

Chinese Version

ZHANG Long
August, 2020

Contents

Module 1 Technology Foundations of Machining Center 1

Task 1 Introduction of CNC Machining Center 1

Task 2 Reading and Plotting of Production Part Drawings 8

Task 3 Basic Techniques Analysis of Machining Center 13

Task 4 Instruction on Measuring Tools in Machining Center 29

Task 5 Basic Mold Fixtures and its Usage Methods 43

Task 6 Installation and Adjustment of Basic Cutting Tools in Machining Center 49

Module 2   Basic Programming in Machining Center 54

Task 1 Geometric Coordinate System of the Machine Tool in Machining Center 54

Task 2 Basic Manual Programming in Machining Center 60

Module 3   Basic Operation of Machining Center 77

Task 1 Use of the Operation Panel of Machining Center 77

Task 2 Basic Operation Requirement of Machining Center 81

Task 3 Inputting and Editing of CNC Machining Program 87

Task 4 Tool Setting of Machining Center 92

Task 5 Debugging and Operation of CNC Machining Program 103

Module 4   Basic Preparation and Inspection of Machining Center 106

Task 1 Preparation on Operation 106

Task 2 Dimensional Accuracy Control in Plain Milling 111

Module 5   Basic Maintenance of Machining Center 119

Task 1 Maintenance Routine of Machining Center 119

Task 2 Maintenance System of Machining Center 121

Module 1

Technology Foundations of Machining Center

Task 1　Introduction of CNC Machining Center

【Knowledge objectives】

(1) Understand the basics of Computer Numerical Control (CNC) technology;

(2) Understand the structure of CNC machining center;

(3) Understand the operating procedures, professional norms and production safety requirements.

【Ability objectives】

Be able to operate the machining center in accordance with its operating procedures and occupational specifications.

1. An Introduction of CNC Technology

1) Definition of CNC machine tool

CNC Machine Tool, also known as CCM(Computer numerical Control Machine tool), refers to automated machine tool equipped with the program control system (PLC). The PLC is able to carry out logic processing of the programs prescribed by control codes or other symbolic instructions. To be specific, with PLC, these control codes and symbolic instructions can be encoded and represented as digital codes and then input into CNC device through information carrier. By computing these digital codes, the CNC device sends various control signals to control the movement of the machine tool, so as to automatically process workpiece with the identical shape and size specified in drawings. As a highly flexible and efficient machine system, CNC machine tool provides good solutions to the processing of machine parts with characteristics of complexity, precision, small-batch and multi-type. It is a typical product of electromechanical integration and leads the development of modern machine tool control technology, a picture of CNC machine tool is shown as in Figure 1-1-1.

Generally speaking, CNC machining refers to the processing procedures of machine parts on CNC machine tool. CNC machining technology refers to the theories, methods and skills for the high-efficiency and high-quality processing of machine parts, especially complex parts. It is the fundamental and key technology for the automatic, flexible, agile and digital manufacturing, which integrates the traditional machinery manufacturing, computing, modern control, sensor detection, information processing and optical-electromechanical technology. Therefore, it is also regarded as the basis of modern machinery manufacturing technology. The wide application of this technology has reformed the production mode and product structure of machinery manufacturing industry. Today, the development and popularity of CNC technology has become one of the key parameters representing the comprehensive strength and industrial modernization level of a country.

Figure 1-1-1   CNC Machine Tool

2) Development of CNC Machine Tool

In the mid-20th century, the development of electronic technology, automated information and data processing as well as computer technology brought new concepts to automation technology. The use of digital signals to control the movement and processing procedures of machine tools promoted the development of machine tool automation. In 1952, MIT installed an experimental numerical control system on a vertical milling machine and used it to successfully control simultaneous three-axis movement. Since then, it has been regarded as the first CNC machine tool in the world.

Among various types of CNC machine tools, the machining center is the most noteworthy one, it refers to a machine tool used as an automatic cutting tool changing device and processes multi-processing with once-a-time clamping of workpieces. A machining center was originally developed by Keaney & Trecker Corp. in March, 1959. The tool bank of this machining center contains screw taps, drills, reamers, milling cutters and other tools. During the processing, the machining center can automatically select the cutting tools under the instructions of punched tape and mount the cutting tools on spindles by manipulators. Therefore, the time of workpiece loading and unloading as well as the replacement of cutting tools can be greatly shortened. Machining center has become a very important type among CNC machine tools. Various types of machining center have been developed, including vertical and horizontal boring-milling centers for box parts processing, and turning and grinding centers for revolving monolithic parts processing. At machining center, once a workpiece is clamped, more than two faces of the workpiece can be simultaneously processed. In addition, selection and change of various functions of cutting tools can significantly improve the processing efficiency of machining center.

In terms of different processing procedures, machining centers can be grouped into two categories, boring-milling machining center and turning machining center; according to the number of control shafts, they can be divided into three categories, three-axis machining center, four-axis machining center and five-axis machining center.

3) Characteristics of CNC Machine Tool

Machining center is the brain of CNC machine tool, it is used to operating and monitoring the machine tool. Compared with other general machine tool, CNC machine tool has the following characteristics.

(1) High flexibility. The processing of parts on CNC machine tool depends mainly on processing procedures. Different from other general machine tools, there is no need to produce and replace various molds and fixtures and to adjust machine tool frequently for the processing of parts on CNC machine tool. Therefore, CNC machine tool is suitable for the processing of the parts which are frequently changed. In other words, it is suitable for the processing and production of single-piece, small-batch and newly-developed products, because it can greatly save production preparation cycle and costs on equipment.

(2) High processing accuracy. The processing accuracy of CNC machine tool can reach 0.005−0.01 mm. CNC machine tool is controlled by digital signals. Each pulse signal output from the numerical controller enables the moving component of the machine tool to move in a pulse equivalent distance (usually 0.001 mm). Moreover, the backlash of the feed drive chain and the average error of the screw pitch can be decreased through curvature compensation of the CNC device. Therefore, CNC machine tool is relatively high in positioning accuracy.

(3) Stable and reliable processing quality. The same batch of parts can be processed on the same CNC machine tool, under the same processing conditions, there is no need to change cutting tools and processing procedures, thus guaranteeing the consistency and stable quality of the processed parts.

(4) High efficiency. CNC machine tool can effectively reduce workpiece processing and non-cutting time. High spindle speed and large feed range of CNC machine tool enables heavy cutting. CNC machine tool is entering an era of high-speed machining in which, with the rapid movement and accurate positioning of moving components as well as high-speed machining, the processing efficiency of CNC machine tools can be significantly elevated. In addition, with the help of the cutters in the tool bank of the machining center, the continuous multiple processing can be performed on one machine tool, which reduces the turnaround time during the procedures of semi-finished products processing, and improves processing efficiency.

(5) High degree of automation. CNC machine tool can reduce labor intensity and improve working conditions. High automation can make the processing less laborious. Before the processing, the CNC machine tool is adjusted, and processing program is input. After that, the machine tool is started to automatically and continuously process to get the finished parts. What

the operator needs to do is just input and edit the program, load and unload the parts, prepare cutting tools, monitor the processing and inspect the processed parts. Therefore, the labor intensity is greatly reduced. Meanwhile, parts processing, a traditional labor-intensive work, becomes a modern intelligence-intensive work.

(6) Modern production management. CNC machine tool can accurately estimate the processing time in advance, and the cutting tools and fixtures can be managed in a standardized and modernized way, which makes the standardization of processing information. Having been effectively combined with the Computer Aided Design/Computer Aided Manufacturing (CAD/CAM), CNC machine tool now serves as the basis of modern integrated manufacturing technology.

2. Details of CNC Machining Center Tool

1) Basic composition of VMC850B CNC machining center

The basic composition of VMC850B CNC machining center is shown in Figure 1-1-2.

Figure 1-1-2   VMC850B CNC machining center

The machining center consists of two main parts, the host part and the control part.

The host part is the mechanical structure of the machining center, including lathe bed, spindle box, worktable, base, column, cross beam, feed mechanism, cutter bank, tool changer and auxiliary systems (gas-liquid, lubrication, cooling, etc.).

The control part consists of hardware and software. Hardware includes Computer Numerical Control (CNC) device, Programmable Logic Controller (PLC), output and input equipment, spindle drive device and display device.

2) Basic parameters for VMC850 CNC machining center

The spindle drive system of VMC850B CNC machining center is capable of stepless speed regulation and constant linear velocity cutting. Under the control of numerical control system, $X/Y/Z$ axes can interact with each other in a coordinated way. The tool bank contains 20 cutters. CNC machining center owns a strong comprehensive processing ability. For example, CNC machining center can perform multiple processing procedures to once-a-time clamped workpieces, at the same time, it can achieve high processing accuracy. In terms of the processing of the batch workpieces in medium difficulty, its efficiency is five to ten times higher than that of other general equipment. Particularly, CNC machining center can be used to process the workpieces that other general equipment is incapable to process, therefore, it is suitable for the processing of single part of complex shape and high precision requirement or for the processing of small and medium batch multiple types workpieces.

Main specifications and technical parameters are listed below.

① Worktable size: 500 mm × 1050 mm.

② Maximum worktable load: 600 kg.

③ Spindle taper: BT40.

④ Spindle motor power: 7.5/11 kW.

⑤ Maximum spindle speed: 8000 r/min.

⑥ Maximum feed speed (X/Y/Z): 10 000 mm/min.

⑦ Left and right travel (in X axis): 800 mm.

⑧ Forward and backward travel (in Y axis): 500 mm.

⑨ Upper and bottom travel (in Z axis): 550 mm.

⑩ Net weight: 5000 kg.

3) Characteristics of machining center

Machining center refers to a highly electromechanical integrated machine tool. After workpieces being clamped, the numerical control system can control the machine tool to automatically select, change and align cutters, and change spindle speed and feed rate. Procedures like drilling, boring, milling, reaming, and tapping can be continuously performed on it, greatly reducing the time of workpiece clamping and the auxiliary time of measurement and machine adjustment. It is particularly suitable for the processing of the parts of complex shape, high precision and various types. High precision machining center can replace precision coordinate boring machines, more importantly, it can serve as the basis of the flexible manufacturing cells and systems.

Machining center is suitable for the processing of single workpiece in complex shape and high precision requirement or for the processing of small and medium batch of multiple types workpieces. Particularly, machining center can replace the tooling and special-purpose machines to meet the same product quality and efficiency standards. Therefore, the application of machining center can significantly save the time and cost for the development and upgrading of new products, which ultimately improve enterprise competitiveness.

Machining center is one of the most productive and widely-used CNC machine tools in the world.

It should be noted that machining center and processing unit are different. Machining center refers to a CNC machine tool for processing and manufacturing parts, while processing unit is a processing area composed of multiple machine tools for the main function of processing the delivered parts.

4) Working procedures of machining center

The working procedures of the machining center includes the following:

① Make workpiece processing plans according to drawings of workpieces, and process programming of workpiece processing procedures manually or automatically by computer.

② Transform various actions of machine tools and all the procedure parameters for workpiece processing into information codes recognizable to numerical control device.

③ Store the codes from the information carrier such as CF (Compact Flash) card and USB drive.

④ Insert the information carrier into inputting device which reads the information and input, and then transfer them into CNC device by DNC (Direct Numerical Control) inputting mode.

⑤ Save the workpiece program in the upper level computer which continuously outputs the program to CNC system to perform the processing.

What is introduced above is the most commonly used working procedures of machining center.

Figure 1-1-3 shows the working principle of machining center.

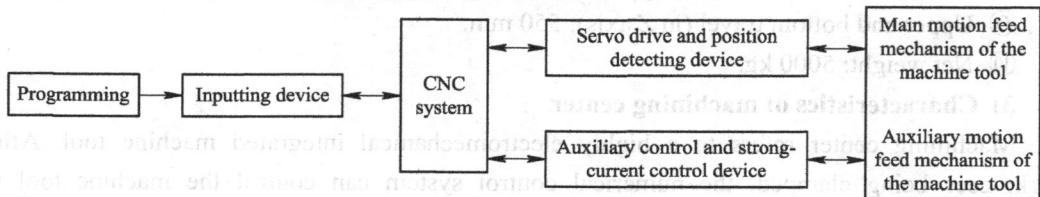

Figure 1-1-3   Working Principle of Machining Center

Another programming and inputting method is to input and output the program directly through the interface between computer and machining center. The program information input into numerical control device is transformed into pulse signals through a series of processing and computing operations. Some signals are sent to the servo system of the machine tool which is used to transform and amplify the signals to drive, through the transmission mechanism, the relevant components (including cutters) of the machine tool and the workpieces operate and move strictly as programmed. Some other signals are sent to programmable controller to sequentially control other auxiliary actions of the machine tool so as to change the cutters automatically.

3. Operation Codes and Professional Norms for Machining Center

The following safety operation instructions should be conformed strictly when we use the machining center.

① Before operating the machine tool, uniforms must be dressed tightly. Operators must wear helmets. Wearing gloves are strictly prohibited during operation.

② After powering up, check the machine tool for any anomalies.

③ Cutting tools (cutters) should be well padded, aligned and tightly clamped; loaded workpieces should be well adjusted and tightly clamped; chuck spanner should be removed immediately after we finish the loading work.

④ In the process of manually moving the carriage or aligning the cutters, feed rate should be slow when the cutter tip goes close to the workpiece, and make sure to press the "Shift" button for position change. Attention must be paid to not press the "Cutter Change" button and the "Shift" button at same time.

⑤ Without being simulated or checked by the instructor before starting the processing, the program should not be used to automatically process the workpieces.

⑥ Before the automatic processing, make sure that the coordinates of the cutter starting point are correct; the protective door of the machine tool should be closed and not be opened at will during the whole processing time.

⑦ The operator should monitor the whole processing operation although it is in automatic processing.

⑧ In any case of abnormality, press the "Emergency Stop" button immediately and report the case in time for cause analysis.

⑨ Any deletion or edition of the program in operation should be strictly prohibited.

⑩ Any change in parameter setting of the machine tool should be prohibited.

⑪ Remove chips with hooks instead of hands; stop the machine to remove the chips in case they are in contact with workpieces; do not put anything on the machine surface during automatic processing.

⑫ Before shut down the machine, make sure the spindle of machine tool be put in appropriate area.

4. Skills Training

Complete the task sheet shown as in Table 1-1-1 based on what you have learned above.

Table 1-1-1　Task Sheet

Item	Safety Operation Codes for Machining Center		
Class		Group No.	
Tasks			
What is the basic composition of VMC850B?			
What are the safety operation instructions for machining center?			

Task 2　Reading and Plotting of Production Part Drawings

【Knowledge objectives】

(1) Learn to recognize part drawings.
(2) Learn to read part drawings.
(3) Learn to make simple part drawings by hand.

【Ability objectives】

(1) Grasp the methods of drawings reading.
(2) Grasp drawing methods and steps.

1. Elements of Part Drawings

Part drawings, also known as the production part drawings, are graphics on which the shape, size and characteristics of a single part are specified and, it is used for the production and inspection reference of the parts. The preparation, processing and inspection of parts are all carried out on the basis of the part drawings and the technical specifications. Therefore, part drawings sheet are important technical document to guide the production of parts.

In order to meet the production requirements, an example of a part drawing sheet is shown as in Figure 1-2-1.

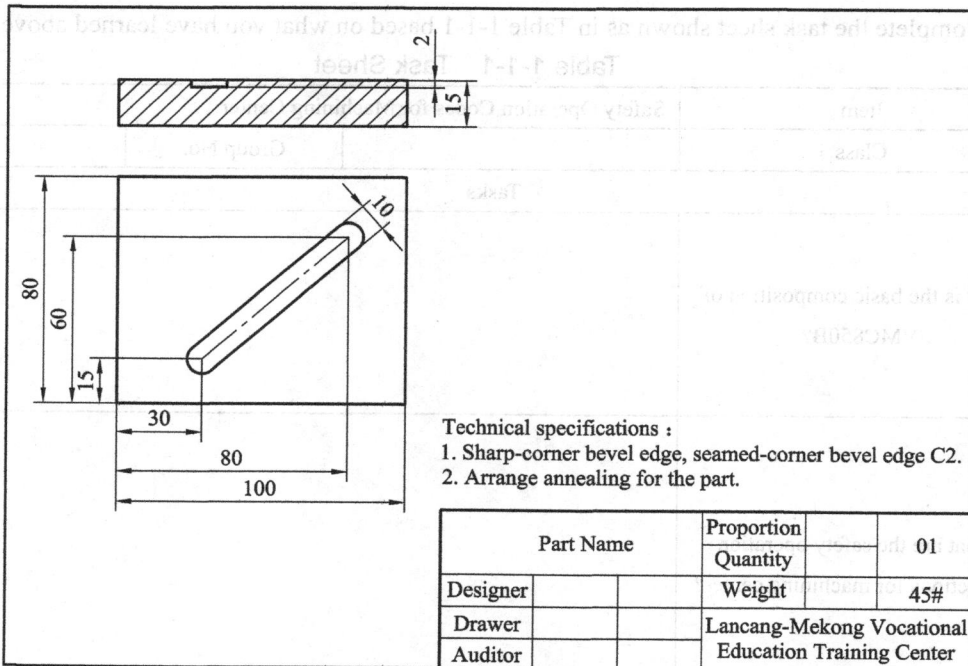

Figure 1-2-1　Part drawing Sheet

A complete part (workpiece) should include the following basic elements:

(1) A group of views. Views should be able to clearly express the internal and external structural shapes of the parts, it can be selected by partial, sectional view, and specifications for simplified drawing.

(2) Size parameters. Size parameters is used to determine the sizes and positions of various components of processed parts. All dimensions required to complete the processing and check whether the parts are qualified for the indications of the parts drawings. Graphics on parts drawings only illustrate the shape of parts, while the real sizes and positions of the various components of the parts are determined by the size parameters on the drawings. The size parameters marked on the part drawings should meet not only the design requirements, but also the production requirements. Therefore, size parameters on part drawings should be complete, clear and in line with relevant national standards.

(3) Title bar. The title bar includes the information about the part name, material, quantity, drawing date, drawing number, proportion, signatures of the drawer and auditor and so on. The form of the title bar should be in line with the relevant national standards. For example, it should be located in the lower right corner and contains the relevant information including the drawer, auditor, material and the part name.

(4) Technical specifications. List the specifications for the production, inspection and use of the parts, including surface roughness, dimension tolerance, shape and position tolerance, material treatment and surface treatment, with symbols, numbers, letters and text annotations in a brief and accurate manner.

2. Reading of a Part Drawing

Below are the procedures for reading part drawing.

(1) Read the title bar. Get to know the name, material, weight of the parts and the proportion of the drawing.

(2) Imagine the shape through an analysis of views. The internal and external shape and structure of parts are key information that should be obtained through drawing reading. The reading methods (verified from the partial view, sectional view, profile view, etc.) are same for assembly drawings and part drawings.

It is necessary to see the basic internal and external structures of the parts and obtain the shapes of some components or oblique planes of the parts from partial view, oblique view and sectional view. In addition, it is important to understand the functions of some structures of the part from design and processing specifications listed on the drawing.

(3) Analyze dimensions and technical requirements. Understand the shaping, positioning and overall dimension of the part as well as the reference used to mark the dimension. In addition, technical requirements, such as surface roughness, tolerance and fits, should also be in mind through drawing reading.

(4) Consider comprehensively. Integrate the structural shape, dimension parameters and

technical requirements for fully understanding of the part drawing.

Sometimes, relevant technical materials, such as sub-assembly drawings and other related components drawings, should be referred to understand complex part drawings.

Read the part drawing of the mounting plate shown as in Figure 1-2-2.

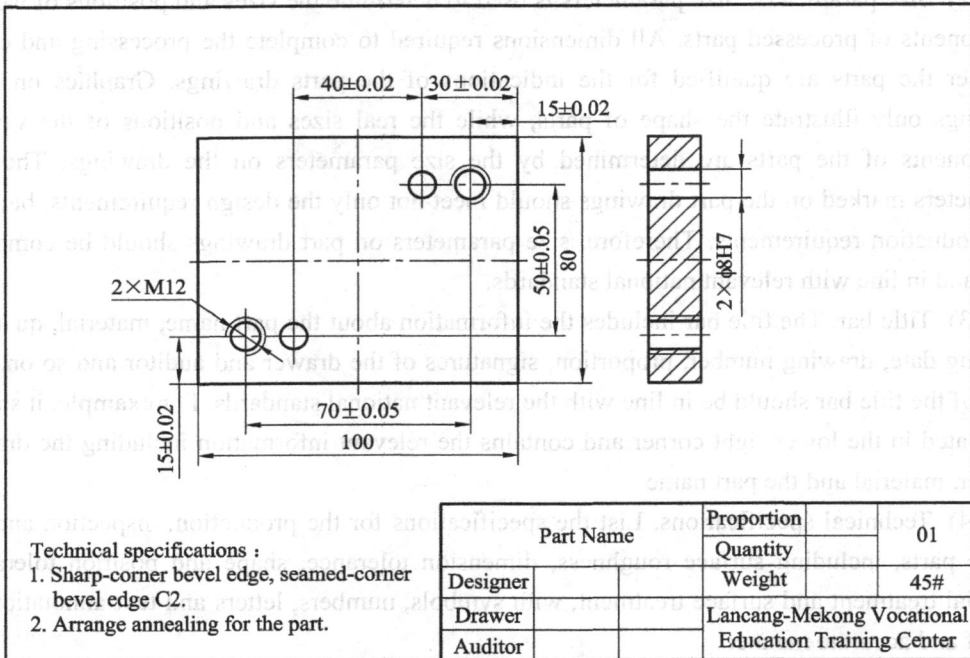

	Part Name		Proportion	01
Technical specifications :			Quantity	
1. Sharp-corner bevel edge, seamed-corner bevel edge C2.	Designer		Weight	45#
	Drawer		Lancang-Mekong Vocational Education Training Center	
2. Arrange annealing for the part.	Auditor			

Figure 1-2-2   Part Drawing of Mounting Plate

3. Simple Part Drawings by Hand

Commonly-used hand drawing tools are drawing plate, T-ruler, triangle board, compass, sub-ruler, scale, curve board, pencil, drawing paper and other drawing tools.

1) Procedures for hand drawing

Understand and Follow the drawing standards, correctly using drawing tools and grasping geometric drawing methods, in addition to that, the drawer should also master proper drawing sequences so as to ensure the drawing quality and improve the drawing speed. The methods and steps for manual drawing with drawing tools are described as follows.

(1) Preparations before drawing. Before drawing, prepare drawing tools and instruments as well as drawing papers; sharpen the lead of pencils according to the required line shapes.

(2) Draft drawing procedures to make draft drawing are as follows.

① Set proper scale and sheet area of the drawing according to parts dimensions.

② Fix the drawing paper on the drawing board with paper tape. In fixing the drawing paper, parallel the horizontal edge of the paper to the long edge of the T-ruler, and keep the distance between the bottom edge of the drawing paper and the bottom edge of the drawing board greater than the width of the T-ruler.

③ Draw the boundary lines, frame and title bar according to the related standards of the width, surrounding size and title bar position of the drawings.

④ Figures should be located in proper area of the drawing paper instead of being excessively close to the boundary lines or any corner. When in drawing the datum line of each view, leave enough space for the description of dimensions and notes based on the length and width of each view.

⑤ Draw the datum lines and outlines of the view on the basis of the positioning and shaping dimensions, respectively, and then draw the details of the view. A hard pencil (2H or H) is recommended to draw drafts. Lines of drafts should be light, fine and accurate so as to erase and modify the draft easily.

(3) Darken the pencil lines. Before darkening the drawing lines, drafts should be carefully checked for correcting errors, erasing redundant lines or stains. Make sure that the lines conform to the related standards as follows:

It should be noted that different types of thick and solid lines should be darkened first, then the thin and dotted lines; for curve and straight lines, curve lines should be darkened first; for multiple horizontal lines, upper lines should be darkened first; for multiple vertical lines, lines on the left should be darkened first; for multiple concentric circles, the smaller ones should be darkened first; the oblique lines, border lines and title bar should be darkened at end.

Different types of lines should be darkened with different types of pencils.

(4) Mark dimensions. After the lines being darkened, the extension lines, dimension lines and arrows should be drawn at once, and then, dimension number and drawing symbols should be noted. The dimension number must be accurate, clear and in line with related standards.

(5) Fill in the title bar and other necessary instructions. In engineering drawings, the title bar should be arranged for the convenience of reading and accessing related information. Generally, the title bar is placed in the lower right-hand corner of the drawing sheet and its direction should be consistent with the direction in which to read drawings.

(6) Check the drawing. Check the drawing carefully to make sure that there is no error and omission after all the drawing work is completed. Finally, sign your name and drawing date in the "drawer" column.

2) Basic requirements for hand drawing

① The lines should be smooth and clear.

② Try to make the visual measurement of the dimensions as accurate as possible (try to be close to the real dimensions) and set proper proportions of the different parts of the drawing. Pencil can be used as ruler to approximately measure the dimensions of an object for the draft drawing, the use of pencil as ruler is as shown in Figure 1-2-3. In addition, relative proportions among different segments of the parts should be estimated for a reduced size part drawing.

③ Keep fast in drawing speed.

④ Legends and dimensions should be written and marked in a neat and correct manner.

⑤ The pencil used for hand drawing should be in one degree softer than the one used for instrumental drawing. The pencil head should be sharpened in a conical shape. The pencil for drawing thin lines should be sharper than that for drawing thick lines.

Figure 1-2-3　The Use of Pencil as Ruler in Hand Drawing

4. Skills Training

Complete the task sheet as in Table 1-2-1 based on what you have been learned above.

Table 1-2-1　Task Sheet

Item	Hand drawing of linear groove parts		
Class		Group No.	
Tasks			
Drawing paper	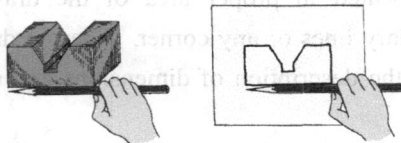 Technical specifications : 1. Sharp-corner bevel edge, seamed-corner bevel edge C2. 2. Arrange annealing for the part.		
Hand drawing			

Part Name table:

Part Name		Proportion	01
		Quantity	
Designer		Weight	45#
Drawer		Lancang-Mekong Vocational Education Training Center	
Auditor			

Task 3　Basic Techniques Analysis of Machining Center

【Knowledge objectives】

(1) Master technical characteristics of machining center.

(2) Analyze techniques of machining center and determine processing plans.

(3) Master basic technical designs for machining center.

【Ability objectives】

(1) Master the methods for techniques analysis in programming.

(2) Master the methods for processing plan design.

1. Technical Characteristics of Machining Center

1) Main processing objects of machining center

The machining center is suitable for processing parts of complex shapes by multiple procedures and high precision requirements. The clamping and adjustments require multiple types of general machine tools with many cutters and fixtures for many times. Therefore, its main processing objects include five types: box parts, complex surface parts, irregular-shaped parts, disk, sleeve and plate parts, and special processing parts.

(1) Box parts. Box parts generally refer to the parts that have internal cavities and more than one hole-systems as well as certain proportions in the length, width and height directions. Such parts are widely used in machine tool manufacturing, automobile manufacturing, aircraft manufacturing and other industries. Machine spindle box, gearbox, engine case cover and engine crankcase are commonly used box parts. Figure 1-3-1 shows some typical box-type parts.

(a) Machine spindle box

(b) Gearbox

(c) Engine case cover

(d) Engine crankcase

Figure 1-3-1　Several typical box parts

Below is the processing techniques analysis of box parts.

Box parts usually require multiple hole-system with plane processing, tighter processing tolerance, and especially the geometric tolerance. The processing generally includes milling, drilling, expanding, boring, reaming, countersinking, tapping and other procedures which apply many sets of tools and cutters. In addition, there are also problems of high costs, long processing cycles, clamping choosing and adjustments, measurements by hand as well as frequent cutters changes. Therefore, it is difficult to stabilize the processing techniques or ensure the processing accuracy while processing the box-type parts on general machine tools.

Horizontal boring and milling machining centers are usually used to process the box parts requiring multiple processing stations and angle rotations, while vertical machining centers are used for the box-type parts requiring fewer processing stations and small processing spans (it means that the processing can be started from one end of the parts).

In small and medium batch production of box parts, the processing route is usually as follows: blank casting → aging → painting → marking → roughing and finishing datum plane → roughing and semi-finishing various planes → roughing and finishing main holes → roughing and finishing secondary holes (thread holes, fastening holes and oil holes)→ deburring → cleaning → inspection.

In mass production of box parts, the processing route is usually as follows: blank casting→ aging → painting → roughing and semi-finishing datum plane → roughing and semi-finishing various planes → finishing datum plane → roughing and semi-finishing main holes → finishing main holes → roughing and finishing secondary holes (thread holes, fastening holes, oil holes and through-holes) →finishing various planes → deburring → cleaning → inspection.

(2) Complex surface parts. Complex surface parts play an important role in mechanical manufacturing industry, especially in aerospace industry. It is difficult or even impossible to process the complex surface parts by the general machining method. The traditional method for processing the complex surface parts — precision casting is relatively low in precision.

The complex surface parts usually include impellers, wind guide wheels, spherical surfaces, various curved surface molds, screw propellers, underwater vehicle propellers as well as some other free-form surfaces. Complex surface parts can be divided into four categories.

① Cams. As the basic components for mechanical information storage and transmission, cams are widely used in various automatic machines. There are different types of cams, including disc cams, cylindrical cams, conical cams, bucket cams, end cams, etc. (See Figure 1-3-2.) Three-axis, four-axis or five-axis machining centers can be selected to process cams in terms of the complexity of each of them in structure.

② Integral impellers. Integral impellers are commonly used in compressors of aircraft engines, expanders of oxygen producing equipment and single screw air compressor, etc. (See Figure 1-3-3.) The machining centers with more than four axes can be used to process such parts.

(a) Disc cams   (b) Conical cams   (c) End cams

Figure 1-3-2   Cams

Figure 1-3-3   Integral Impellers

③ Casting molds. There are many types of casting molds, such as injection molds, rubber molds, vacuum forming molds, refrigerator foam molds, pressure casting molds, precision casting molds, etc. Because of the highly centralized working procedures, precision casting molds, such as dynamic and static molds, is basically used in one clamping, which can reduce the cumulative dimension error and the workload of repairing. At the same time, the molds processed by machining centers are strong in reproducibility and good in interchangeability. In addition, the processing of casting molds by machining tools can greatly decrease fitters' workload, because machining tools can process the parts as much as they can and only leave polishing to be done by the fitters. Shown in Figure 1-3-4 are two sections of bath mold.

Figure 1-3-4   Sections of Bath Mold

④ Spherical parts. Spherical parts are suitable for milling by machining centers. On three-axis milling centers, spherical parts can only be approximated and processed by ball-end milling cutters, low in efficiency. However, on five-axis milling centers, end milling cutters can form an envelope surface for processing the spherical surface approximation. That programming workload is relatively large for processing complex surface such as spherical surface by machining centers, requires to apply automatic programming technology.

(3) Irregular-shaped parts. Irregular-shaped parts refer to the parts that are of irregular outlines and require multi-station combined processing at different points, lines and surfaces. There are some limitations of the irregular-shaped parts processing, such as poor in rigidity, easy to deform under pressure, low in processing accuracy and even complex for some subparts to be processed on general machine tools. However, all these limitations can be overcome by using machining centers for their reasonable processing techniques, single or multiple clamping, combining processing at points, lines and surfaces for multiple or even combining all the processing procedures. Figure 1-3-5 shows some typical special-shaped parts.

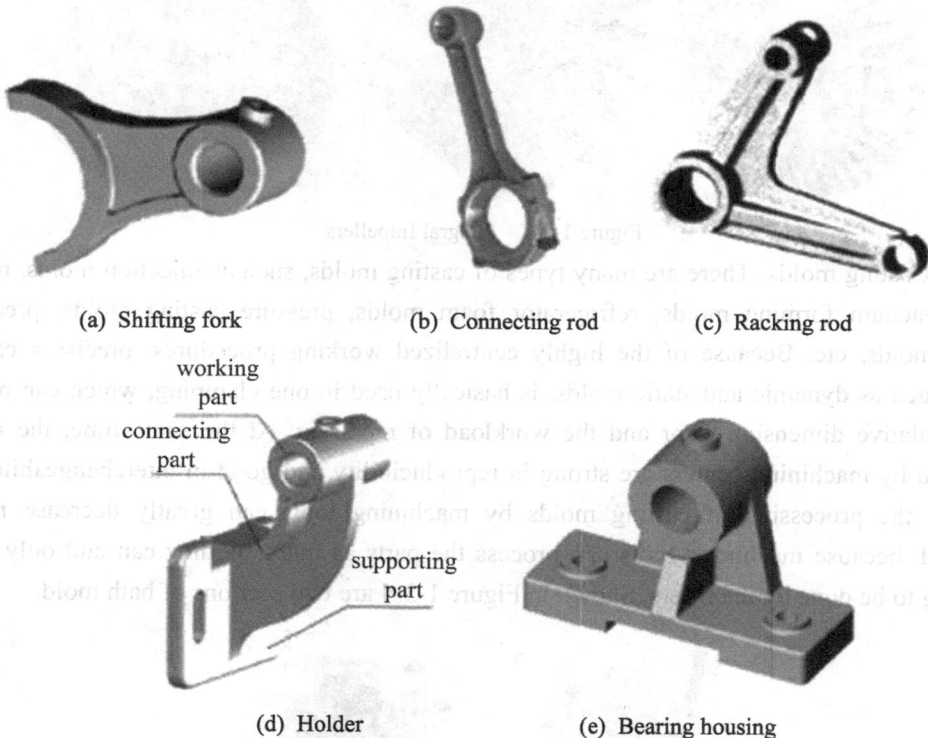

(a) Shifting fork   (b) Connecting rod   (c) Racking rod

(d) Holder   (e) Bearing housing

Figure 1-3-5 Special-shaped Parts

(4) Disk, sleeve and plate parts. There are different types of disk, sleeve and plate parts. There are curved disk parts and sleeve parts including axis sleeves with flanges and shaft parts with keyway or square heads. There are plate parts with multiple holes, such as motor covers, as shown in Figure 1-3-6. Some disk parts, sleeve parts and plate parts contain axis holes, flanges, bosses and pits; some contain multiple thread holes, light holes, countersunk holes, pin holes and keyway; others contain wheel spokes, spoke plates, rib plates and sealing structures such as oil

grooves and felt ring grooves.

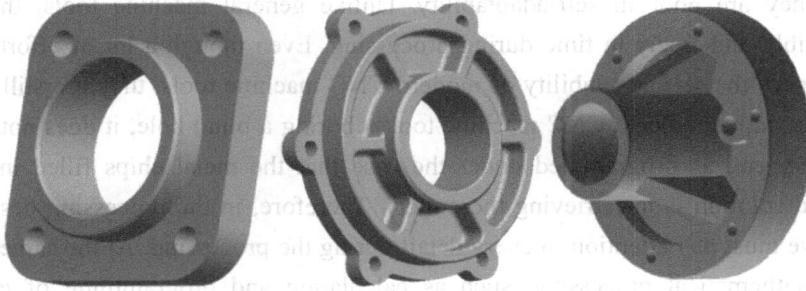

Figure 1-3-6   Disk, Sleeve and Plate Parts

Disk parts often require high flatness and axial dimension accuracy at the supporting end face and parallelism of the two ends; in addition, they also have a demand on high verticality of the inner holes for switching and on coaxiality between the outer circle and inner holes. Vertical machining centers are suitable for the disk parts with distributed holes or curved-end surfaces, while horizontal machining centers are suitable for those of radial holes.

(5) Special processing parts. Having mastered the functions of the machining centers and supplied with certain constriction suit and special tools, we can perform some special processing, such as letters engraving, lines marking and metal surfaces patterns carving. By connecting the spindle of the processing center with a high-frequency power, the metal surface can be quenched by line scanning, and by installing a high-speed grinding head to the spindle of the processing center, little modulus involute bevel gears with various carves on metal surface can be made.

2) CNC processing technology

CNC processing technology is the sum of the methods and techniques used in parts processing with CNC machine tools.

In many aspects, the principle of CNC machining is basically the same as that of the general machine tool processing. Whereas, CNC machine tools is in highly automated, different in control methods and high in cost, CNC processing technology has its own characteristics accordingly as follows.

(1) Specific regulations design. When processing parts by general machine tools, many specific technological questions, such as the division and sequence of processing steps, the geometric shape of the tools, the path of the tools and the cutting parameters, are largely determined by the operators according to each of their practical experience and habits, and in general, there is no need for the technicians to design and follow detailed regulations in technological procedures. However, in CNC machining, not only must these specific technological problems be considered seriously in designing processing procedures, but also must these specific technologies be correctly selected and incorporated into the processing program. That is to say, many specific technological matters and details originally flexibly handled by operators through timely adjustments must be designed and arranged by programmers before CNC machining.

(2) Rigorous and integrated technical design. Although CNC machine tools are highly automated, they are poor in self-adaptability. Unlike general machine tools, they can't be adjusted flexibly and freely in time during processing. Even though a lot of efforts have been made to improve the self-adaptability of modern CNC machine tools, they are still not flexible enough. For example, when a CNC machine tool is boring a blind hole, it does not stop boring till to the very end as programmed under the case that the metal chips filled in the hole (a recommended solution is to retrieving the cutter). Therefore, in the processing design of CNC machining, we must pay attention to every detail during the processing. At the same time, in the process of mathematical processing, such as calculation and programming of graphics, we should strive to be accurate enough, so as to ensure smooth processing. In practical work, even a tiny error about a decimal point or a positive or negative sign may lead to severe processing and quality accidents.

(3) Focusing on processing adaptability. When manufacturing parts, correct processing methods and operations should be selected according to the characteristics of CNC machining. High automation, high stability in production quality, simultaneous muti-axis movement, and great convenience in performing multiple processing procedures combined with expensiveness and requirements for great operating skills, make it necessary to require proper selection of processing methods and processing objects to avoid great losses. In order to give full play to the advantages of CNC machining and achieve high economic profits, special care must be taken in selecting processing methods and objects, sometimes, the shape, size and structure of the workpiece should be modified accordingly without changing the original performance of the workpiece in CNC machining.

Generally, in the process of selecting and determining the operations of CNC machining, the technicians must make a detailed and sufficient technical analysis of the parts drawing or parts model. When conducting the analysis of the technical design of CNC machining, programmers should, based on the basic features of CNC machining and functions and actual processing experience of the machine tool, strive to make this preliminary preparation more carefully and solidly, so as to pave the way for the following work, that is, reducing errors and reworking, and leaving no hidden troubles. In other words, the technological design of CNC machining must be completed before programming, because the valid programming must be based on the design. Thus, the quality of the design affects not only the efficiency of the machine tool, but also the quality of the parts. In addition, an inadequate technical design is one of the main causes of CNC machining errors according to the report by analysing a large number of processing cases. Therefore, it is of great importance to conduct a comprehensive analysis of the technical design before programming.

3) Details of techniques design of CNC machining center

① Determine the parts suitable for machining on CNC machine tools and determine the processing procedures.

② Analyze the parts drawings; clarify the processing contents and technical requirements;

determine the processing plan for the parts; formulate the technological plan for CNC machining, such as dividing processing procedures, connecting the CNC machining procedures with the non-CNC machining process, etc.

③ Design processing procedures and working steps. Select the positioning datum of parts, determine the fixture and auxiliary tool schematic and the cutting parameters, etc.

④ Adjust CNC machining program. Select the tool setting point and tool changing point, and determine the tool compensation and the processing route.

⑤ Allocate machining allowance in CNC machining.

⑥ Handle some process instructions of CNC machine tools.

⑦ Process a first piece and fix the problems having occurred in the processing.

⑧ Finalize and file the technical documents of CNC machining. Different CNC machine tools have different technical documents. Generally, the documents for CNC milling machine should include:

a. Programming Task Sheet.

b. CNC Machining Procedure Cards.

c. CNC Machine Tool Adjustment Sheet.

d. CNC Machining Tool Selection Cards.

e. CNC Machining Feed Route.

f. CNC Machining Program List.

Among them, the most important are the CNC Machining Procedure Cards and CNC Machining Tool Selection Cards. The former describes the sequence of CNC machining and the processing objects, and the latter is the basis for the use of tools. As a matter of fact, these technical documents should be standardized for better management. However, there is currently no unified national standard, enterprises generally formulate the above-mentioned documents according to their own actual situations.

2. Technical Analysis of the Machining Center and Determination of Processing Plans

There are many contents involved in the technical analysis of CNC machining. From the aspects of the possibility and convenience of CNC machining, the following points should be taken into consideration.

1) Principles for marking dimensions in part drawings

① The dimensions marked in parts drawings should conform to the characteristics of programming. In the parts drawing for CNC machining, it is better to use the same standard to mark dimensions or directly give the method for labeling ordinate dimensions. This labeling method is convenient not only for programming, but also for coordinating the designing standards, technical standards, inspection standards and the setting and calculation of program zero point.

② The geometric elements that make up the outline of the parts should be sufficient. In automatic programming, it is necessary to define all the geometric elements that form the outline of parts. When analyzing a parts drawing, it is necessary to analyze whether the given conditions of the geometric elements are sufficient. If not, it is impossible to model and program the parts to be processed.

2) The structural manufacture abilities of each part of the workpiece

① It is necessary to guarantee the machining accuracy and dimension tolerance required by the parts.

② It is better to adopt uniform geometric types and dimensions for the internal cavity and the shape of the parts, so as to reduce the number of tool changes as many as possible.

③ The structure design of the parts should ensure that tools with large diameter be used for machining. The milling cutter with large diameter can improve processing efficiency and the quality of surface processing. As shown in Figure 1-3-7, the relation between the milling cutter with large diameter for machining and the lower machined contour surface of the parts and larger the arc of the inner groove. Therefore, the fillet radius R of the inner groove should gradually increase, in other words, it should be greater than $0.2H$ if possible where H is the maximum height of the contour surface of the machined parts, so as to obtain good manufacturability. Generally, the tool radius r is 0.8–0.9 times of the fillet radius R of the inner groove.

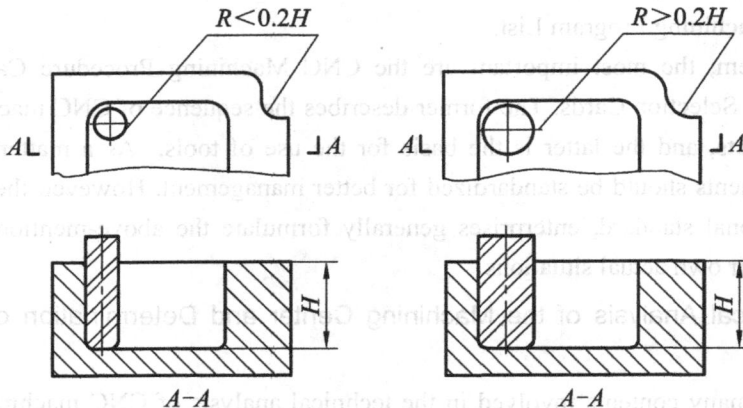

Figure 1-3-7   Comparison of Structural Manufacturability of Inner Grooves (Change in relation between fillet radius of the inner groove and maximum height of the contour surface)

④ The fillet radius R of the groove bottom of the milling surface of the parts or the fillet radius R at the intersection of the web and the margin plate should not be too large. The maximum diameter d of the contact surface between the milling cutter and the milling plane equals $D - 2R$ ($d = D - 2R$, where D is the diameter of the milling cutter). Therefore, when D is fixed, the larger the fillet radius R is, the smaller the plane area being milled by the milling cutter end edge is, and the worse the milling function, efficiency and manufacturability are, impact of groove bottom plane arc of parts on manufacturability can be illustrated as in Figure 1-3-8.

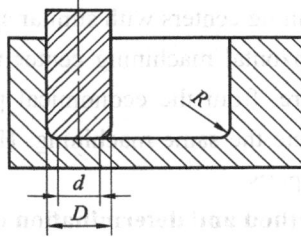

Figure 1-3-8　Parameters Influenced by Inner Grooves

⑤ Uniform datum positioning should be adopted. In the process of CNC machining, if the parts need to be relocated and clamped but there is no a unified positioning datum to follow, there occurs the incongruity of the contour position and of the size of the front and back sides after machining. Therefore, it is necessary to make use of the appropriate holes of the parts in themselves or set up special process holes or take the reference edge of the part contour as the positioning reference to ensure the accuracy of the relative position after two clamping.

3) Selection of CNC milling machine and machining center

There are three types of CNC milling machine, including CNC vertical milling machine, CNC horizontal milling machine, and CNC vertical and horizontal milling machine. CNC vertical milling machine is the most widely-applied type, especially in mold processing. It is commonly used in the manufacturing of small-sized and medium-sized mold, such as plastic injection mold like TV front cover and washing machine panel, die-casting mold like motorcycle cylinder, and forging mold like connecting rod. CNC horizontal milling machine is used mainly for milling planes, grooves and formed surfaces, etc. It is often used for milling mold of deep cavity, such as the cavity of the washing machine barrel and the cavity of refrigerator inner tank. CNC horizontal milling machine usually add CNC rotary tables or universal CNC rotary tables to realize the processing of the fourth axis and the fifth axis. In this way, the machines can not only be used for processing the continuous rotation contour on the side of the workpiece, but also realize "the four-sided machining" by changing the position through the rotatory table in one clamping. The CNC vertical and horizontal milling machine can perform both horizontal and vertical machining on a single machine tool through the change in spindle direction, so as to realize "the five-sided machining", it has the advantage of wider range of applications, but with slightly lower accuracy and rigidity.

The machining center in this book refers to the boring and milling machining center. It combines the functions of milling, drilling, reaming, boring, tapping and thread cutting on one machine, so that it has multiple processing means. After the workpiece is clamped once, it can perform automatic machining on more than two surfaces. Moreover, it has a variety of functions, such as tool selection, tool change and automatic worktable exchange.

There are mainly two types of machining centers, horizontal and vertical machining center. The horizontal machining center is suitable for parts that require multi-station machining and high position accuracy, such as boxes, pumps, valves, housings, etc.; the vertical machining center is suitable for parts that require single-station machining, such as box cover, end cover plane cam, etc. Generally, for machining centers with similar specification (referring to the width of the worktable), the price of horizontal machining center is 50%–100% higher than that of vertical machining center. Therefore, from the economical point of view, vertical machining center should be selected to complete the same machining. However, the horizontal machining center can process a wider range of parts.

4) Selection of processing method and determination of processing plan

(1) Selection of processing method. There are generally two situations for CNC processing. First, according to the existing part drawings and blanks, choose a CNC machine tool suitable for processing the parts. Second, according to the existing CNC machine tool, the suitable parts for processing should be selected. In either case, the appropriate CNC machine tool and processing method should be selected according to the type of parts and the processing contents.

The contour of flat parts is composed mostly of straight lines, arcs and curves, which are generally processed by two-axis CNC milling machine. The parts of three-dimensional surface are mostly processed by three-axis or more-than-three-axis CNC milling machine or machining center. Through rough milling, the dimensional accuracy of the plane can reach IT12–IT14 (referring to the size between the two planes), and the surface roughness value Ra can reach 12.5–50 μm. Through rough and finish milling, the dimensional accuracy can reach IT7–IT9, and the surface roughness value Ra can reach 1.6–3.2 μm.

There are many methods for hole processing, such as drilling, expanding, reaming and boring. Large-diameter holes can also be milled by circular interpolation.

For processing the rough holes of diameters larger than Φ30 mm and which have been cast or forged, the following processing plan is generally adopted, the rough boring — semi-precision boring — hole chamfering — fine boring.

If the hole diameter is large, rough milling and finish milling can be adopted. When there is an undercut groove, the saw blade milling cutter can be used to complete the milling after semi-precision boring and before fine boring. It can also be bored by boring cutter with single cutting edge, which is less efficient.

For processing holes without blank holes and of diameters less than Φ30 mm, the following plan is usually adopted, flat end face countersinking — center hole punching — drilling — expanding — hole chamfering — reaming.

For small holes with coaxiality requirements, the processing plan should be as follows, flat end face countersinking — center hole punching — drilling — semi-precision boring — hole chamfering — fine boring (or reaming). In order to improve the position accuracy of the hole, it is necessary to arrange the step of end face countersinking and center hole punching before the drilling. The chamfering of the hole is arranged after semi-precision fine boring, aka finishing and before fine boring to prevent burrs in the hole.

The method for processing internal thread depends on the size of the hole diameter. In general, the thread of a nominal diameter between M6–M20 is usually processed by tapping.

Threads with a nominal diameter less than M6 are tapped by other means after the bottom hole is processed by the machining center. That's because controlling the machining state in tapping the thread by the machining center is impossible, and taps of small diameter can be easily broken. Threads with a nominal diameter above M20 can be milled with thread milling cutters.

The principle for selecting processing method is to ensure the precision and surface roughness of the processed surface. Since there are generally many processing methods to obtain the same level of accuracy and surface roughness, it is necessary to consider the shape, size and heat treatment requirements of the parts in the actual selection. For example, boring, reaming, grinding and other methods can be used to process the holes of IT7 accuracy, while the holes on the box are generally processed by boring or reaming instead of grinding. Generally, reaming is selected for small-size box holes, and boring should be used when the hole diameter is relatively large. In addition, the requirements for production efficiency and economy, as well as the actual conditions of the factory's production equipment, etc. should also be considered.

(2) Determination of processing plan. In determining the processing plan, first of all, the processing method, thus the finishing method, should be chosen to meet the requirements for the surface precision and surface roughness; then the final plan for processing the blank into the required parts can be decided.

In the process of machining, the workpieces can be divided into plane and curved parts according to the surface contour. The bevel contours of the plane parts can be divided into two types, the contour surface of fixed bevel angle or that of variable bevel angle. When merely considering the processing technology, the best processing scheme is to use multi-axis CNC machine tool for contour surface machining, which can enable high production efficiency as well as good quality. However, the average small-sized and medium-sized enterprises could not afford such expensive and high-production-cost machine tools. Therefore, it is necessary to consider using 2.5-axis and 3-axis machine tools for machining.

The ball end milling cutter is usually used to process curved surface parts by the 2.5-axis and 3-axis CNC machine tools. The machining accuracy of the contour surface is mainly ensured by controlling the tool step length and the width of the processing belt. The smaller the tool step length and the width of the processing belt are, the higher the requirement for the processing accuracy would be, and the lower the programming and processing efficiency would be.

As shown in Figure 1-3-9, the radius of the ball head cutter is r; the radius of curvature on the surface of the parts is ρ; the line spacing is S and the residual height of the curved surface after machining is H. Then, the formula for calculating the line spacing is,

$$S = 2\sqrt{H(2r-H)} \times \frac{\rho}{r \pm \rho}$$

In the formula, the sign "+" is taken when the surface of the processed part is convex in the ab

segment, and the sign "−" is taken when it is concave. There are two main ways to select line spacing and step length in the current CAD/CAM system programming. One is to select equal line spacing and equal step length after estimation. No matter how the curved surface changes in the specified area, the tool always cuts with equal line spacing and step length. Since the curvature and unevenness of the curved surface change, by cutting in this way, the residual height of the curved surface is different. Another is the equal residual height method. In programming, firstly determine the residual groove height on the surface of the entire curved surface. Then, the CAD/CAM system automatically calculates the line spacing and step length according to this height. Therefore, by cutting in this way, no matter how the surface changes, the residual height is always equal, but the line spacing S and step length are not coincide.

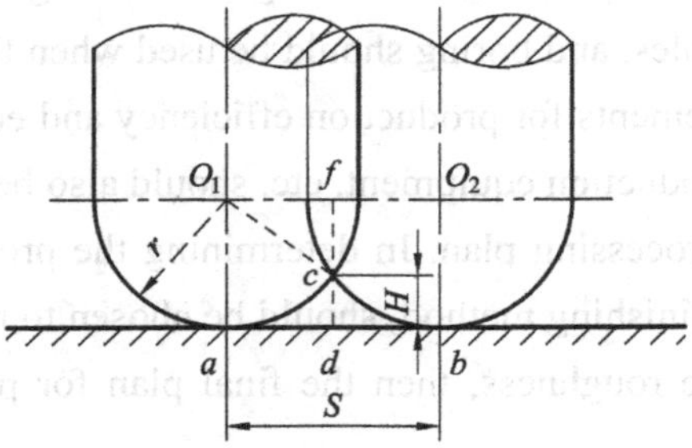

Figure 1-3-9   Calculation of Line Spacing

3. The Basic Technical Design for Machining Center

1) Division of procedures and working steps

When machining parts by CNC machine tools, the working procedures should be concentrated as much as possible, and most of them should be completed in one clamping. There are the following methods for the division of CNC machining procedures.

(1) Divide procedures in terms of processing contents. For parts with much processing contents, the processing contents are divided into several parts according to the structural characteristics of the parts, and each part can be processed with a typical tool. For example, there has to be the processing of inner cavity, outer shape, and plane or curved surface, etc. When machining the inner cavity, the outer shape is in clamping; when processing the outer cavity, the hole of the inner cavity is in clamping.

(2) Divide procedures in terms of the tools. By dividing procedures in terms of the tools, the empty travel can be compressed and the time for and errors of tool changes can be reduced. Although a lot of surfaces of some parts can be processed after one clamping, the program is too long and there come some limitations, such as large memory capacity (mainly on the control system), the continuous working time of the machine tool (such as a procedure cannot be finished in one shift), etc. In addition, a longer program increases the error rate, making error detection and retrieval difficult. Therefore, the program cannot be too long, and the contents of one procedure cannot be too much.

(3) Divide procedures in terms of roughing and finishing. For the parts that are easy to

have deformation, usually they need to be corrected after rough machining. In this case, roughing and finishing are two procedures, that is, roughing first and then finishing, which can be processed with different machine tools or different cutters. To sum up, when dividing the procedures, we must be flexible considering the structure and manufacturability of the parts, the function of the machine tool, the sequence in CNC machining contents of the parts, the sequence in times, and the production and organization capability of the department, etc. Whether the parts are suitable for centralized processing or decentralized processing should be reasonably determined according to the actual needs and production conditions. Moreover, the arrangement for processing sequence should be based on the structure of the parts and the roughing degree, as well as the needs for positioning and clamping. It should be noted that the rigidity of the workpiece can not be damaged. Generally, the processing sequence should be arranged according to the following principles.

① The previous procedure should not affect the positioning and clamping of the next, and the processing procedure inserted with general machine tool processing should also be considered comprehensively.

② The inner cavity processing is carried out first, and then the processing of the outer shape.

③ For the multiple processing procedures in the same clamping, the procedure with less damage to workpiece rigidity should be arranged first.

④ The processing procedure by the same positioning and clamping method or the same tool should be carried out continuously to reduce the times of repeated positioning, tool changing and clamping plate moving.

In order to facilitate the analysis and description of complex procedures, the procedures can be divided into steps further. The division of the steps is considered mainly from two aspects, the processing accuracy and efficiency. For example, if the parts are processed by the machining center, the same surface should be processed in the order of rough machining, semi-finishing machining and finishing machining. That is to say, the entire surface should be processed separately in the sequence from rough machining to finishing machining. For parts that have both milled surfaces and boring holes, the surface can be milled first and then bored, so as to reduce the impact on the accuracy of the hole due to the possible deformation of the parts caused by the large milling cutting force. For the machining center with a rotary table, if the rotation time is shorter than the tool change time, the steps can be divided according to the tool selection to reduce the number of tool changes and improve processing efficiency. However, since the CNC machining is divided according to processing steps, the inspection system (self-inspection, mutual inspection and special inspection) is inconvenient to implement. In order to avoid quality problems of the whole batch of parts, the inspection should be carried out after each step instead of a complete procedure.

2) Selection of machining allowance

The machining allowance refers to the difference between the solid dimension of the blank parts and the drawing dimension of the parts. The size of machining allowance has a great influence on the machining quality and manufacturing economy of parts. An excessive margin not only wastes raw materials and time for machining, but also increases the consumption of machine tools, cutting tools and energy. However, a relatively small margin does not eliminate the various errors, surface defects and clamping errors of the previous procedure, which is likely to cause waste products. Therefore, the machining allowance should be reasonably determined according to the related factors. In general, the processing of parts has to go through rough machining, semi-finishing and finishing to meet the final requirements. Therefore, the total machining allowance of parts is equal to the sum of the machining allowances of the intermediate processes.

First, Principles for selecting machining allowance between procedures are as follows.

① Adopt the principle of minimum machining allowance to shorten the processing time and reduce the processing cost of parts.

② There should be sufficient machining allowance, especially in the final procedure.

Second, In selecting the machining allowance, other conditions should also be considered.

① Due to the different size of the parts, the deformation caused by cutting force and internal stress are also different. The deformation increases if the workpiece gets larger. Thus the machining allowance should be correspondingly larger.

② The parts have deformation during heat treatment, so the machining allowance should be increased appropriately.

③ The machining method, clamping method and rigidity of processing equipment may cause the deformation of parts. Excessive machining allowance causes deformation due to the increased cutting force. All of these factors should be taken into consideration.

Third, Methods for determining machining allowance are as follows.

① Diagram checking. This method is based on the data from the production practices and experimental research at various factories, which are first accumulated into various diagram data and then integrated into manual. When determining the machining allowance, consult these manual, and then make appropriate modifications based on the actual situation of the factory. At present, this method is widely used in factories in China.

② Empirical estimation. This method is based on the actual experience of the technical personnel to determine the machining allowance. In general, in order to prevent to produce waste products due to limited machining allowance, the value of empirical estimation is always larger. This method is often used for small batch production of single-pieces.

③ Analytical calculation. This method is based on certain test data and calculation formula for machining allowance. You can analyze various factors that affect the allowance, and then calculate to determine it. This method is reasonable, but it needs comprehensive and reliable test data. At present, it is used only for processing the very expensive materials and in a limited number of factories which have a need for mass production.

Shown in Table 1-3-1 are the machining allowance chosen by different processing methods.

Table 1-3-1　Machining Allowance of Plane Finish Milling and Grinding

Processing methods	Length of the plane being processed /mm	Width of the plane being processed					
		≤100 mm		>100－300 mm		>300－1000 mm	
		Allowance a	Tolerance (+)	Allowance a	Tolerance (+)	Allowance a	Tolerance (+)
Finish milling	≤100	1.0	0.3	1.5	0.5	2	0.7
	>100－300	1.2	0.4	1.7	0.6	2.2	0.8
	>300－1000	1.5	0.5	2	0.7	2.5	1.0
	>1000－2000	2	0.7	2.5	1.2	3	1.2
Grinding after finish milling (the parts are not calibrated during installation)	≤100	0.3	0.1	0.4	0.12	—	—
	>100－300	0.35	0.11	0.45	0.13	0.5	0.12
	>300－1000	0.4	0.12	0.5	0.15	0.6	0.15
	>1000－2000	0.5	0.15	0.6	0.15	0.7	0.15
Grinding after finish milling (the parts are clamped or calibrated using a micrometer)	≤100	0.2	0.1	0.25	0.12	—	—
	>100－300	0.22	0.11	0.27	0.13	0.3	0.12
	>300－1000	0.25	0.12	0.3	0.15	0.4	0.15
	>1000－2000	0.3	0.15	0.4	0.15	0.4	0.15

Note: 1. In finish milling, the allowance left before the last stroke should be $a \geqslant 0.5$ mm.

　　2. The machining allowance for grinding the heat-treated parts is to multiply the value in the Table 1-3-1 by 1.2.

3) Determination of processing route

In CNC machining, the processing route refers to the trajectory of tool position relative to the workpiece, which is the basis for programming and directly affects the machining quality and efficiency. The following aspects should be considered in determining the processing route.

① Ensure the machining accuracy and surface quality of parts as well as high processing efficiency.

② Reduce programming time and program capacity.

③ Reduce empty cutting time and the cutting stop time to avoid scratching the parts.

④ Reduce the deformation of parts.

⑤ For the processing of hole parts with high position accuracy requirements, it is important to avoid the influence of the machine tool backlash on the position accuracy of holes.

⑥ The processing of complex curved surface parts should be based on the actual shape, accuracy requirements, processing efficiency and other factors to determine whether we should adopt line cutting or ring cutting, or equal distance cutting or equal height cutting.

4. Skills Training

Complete the task sheet as Table 1-3-2 based on what you have been learned above.

Table 1-3-2　Task Sheet

Task name	Technical analysis of processing composite parts		
Class		Group No.	
Task assignment			
Drawing of parts			
Technical analysis			

Task 4   Instruction on Measuring Tools in Machining Center

【Knowledge objectives】

(1) Be able to use common tools in the machining center.

(2) Be able to use common measuring tools in the machining center.

【Ability objectives】

(1) Master the methods for using common tools.

(2) Master the methods for using common measuring tools.

1. Knowing and Using Common Tools in Machining Center

1) Pliers

Pliers are a hand tool used to clamp and fix the workpiece, or to twist, bend and cut the wire. They are V-shaped, usually consisting of handles, fulcrum and jaw.

Generally, pliers are made of carbon structural steel, which is first forged and rolled into the shape of the tongs, then processed by milling, grinding, polishing and other metal cutting procedures, and finally by heat treatment. The picture of a plier are shown in Figure 1-4-1.

Figure 1-4-1   Plier

2) Screwdriver

A screwdriver is a tool used to turn a screw and force it into place. It usually has a thin wedge-shaped head, which can be inserted into the slot or notch of the screw head. A high-quality screwdriver head is made of spring steel of relatively high hardness. A good screwdriver should be hard but not brittle. If the slot of the screw head becomes bald and slippery, the screwdriver can be knocked with a hammer to make the groove of the screw deeper, so that the screw can be unscrewed easily while the screwdriver could remain intact. The screwdriver is often used to pry things, so it is required to be of certain toughness. In general, the hardness of the head of the screwdriver should be greater than HRC60, which is not easy to rust.

There are mainly two kinds of screwdrivers, i.e., straight (minus sign) one and cross (plus sign) one, as shown in Figure 1-4-2. Hex screwdriver is another common one, including the internal hex head and external hex head.

Figure 1-4-2　Screwdrivers

3) Hacksaw

Hacksaw(see Figure 1-4-3) is a common tool for fitters, which can cut workpieces of relatively small size and steel bars like round bars, angle bars, flat bars, etc. A hacksaw consists of a saw frame (commonly known as a saw bow) and a saw blade. The saw blade is installed on the saw frame when in use. Generally, the teeth of the saw blade are facing outward, but if in this way the teeth are easier to be broken, We can install the blade with its teeth facing inward, so as to relieve the damage and prolong its service life. After its use, the saw blade should be removed or the tension nut should be loosened to prevent the deformation of the saw frame and to extend its service life. According to the types of saw blades, one with single-sided teeth and the other with double-sided teeth, hacksaw can also be divided into coarse teeth (14 teeth / 25 mm), medium teeth (18-24 teeth / 25 mm) and fine teeth (32 teeth / 25 mm) ones suitable for sawing different materials. In order to improve the working efficiency and avoid much more damage to the teeth, the fine teeth one is used for cutting hard materials, the coarse teeth one is used for cutting soft materials, and the medium teeth one is used for cutting general materials. The thickness of saw blade is 0.5-0.65 mm, and its width is 10-12 mm. As for the blade length, there are 200 mm, 250 mm and 300 mm. There are two types of saw frames, one with fixed length and the other with adjustable length. The adjustable saw frame has three gears, which are respectively suitable for saw blades of three lengths.

Figure 1-4-3　Hacksaw

4) File

The file is made of carbon tool steel T12 or T13, and through heat treatment and then quenching treatment, its sharpness is improved. There are many fine teeth on its surface and it is a hand tool for filing workpieces. Hand files are often used for micro filing surfaces of metal, wood, leather and others.

The application of files started from very early times and the oldest file that has been found is the bronze file in Egypt around 1500 BC. Modern files are usually made of carbon steel by rolling, forging, annealing, grinding, teeth cutting and quenching. The files are shown in Figure 1-4-4.

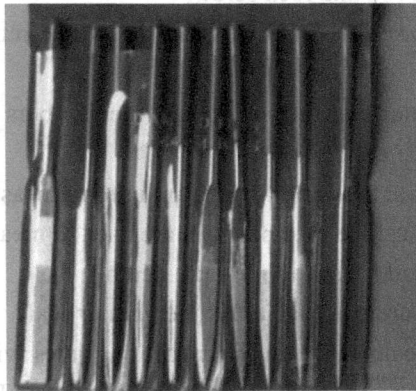

Figure 1-4-4   Files

Below are principles for selecting files.

① Selection of the shape of the blade. The file should be selected according to the shape of the parts being filed, and the shape of the two should match. For filing an internal arc surface, half round file or round file (for a workpiece with a small diameter) should be selected; for filing an internal angle surface, triangular file should be selected; for filing an internal right-angle surface, flat file or square file should be selected. In using a flat file to refine the inner right-angle surface, be careful to make the narrow surface (the smooth edge) of the file without teeth close to a surface of the inner right angle, so as not to damage the right-angle surface.

② Selection of the thicknesses of the teeth of the file. The thickness of the teeth of the file should be selected according to the size of the machining allowance, machining accuracy and material properties. The coarse teeth file is suitable for machining workpieces of large allowance, low dimensional accuracy, large geometrical tolerance, large surface roughness and soft material, otherwise, the fine teeth file should be selected. In being used, the file should be selected according to the machining allowance, dimensional accuracy and surface roughness of the workpiece.

③ Selection of the sizes of the file. The size of the file should be selected according to the size and machining allowance of the workpiece to be processed. If the machining size and the allowance are large, the file of large size should be selected, otherwise, the file of small size should be selected.

④ Selection of the teeth patterns of the file. The teeth patterns of the file should be selected according to the material properties of the workpiece to be filed. For filing aluminum, copper, soft steel and other soft materials, it is better to use single teeth file. Single teeth file is of large front angle, small wedge angle, sharp cutting edge and large chip holding groove which is not easy to block.

2. Knowing and Using Common Measuring Tools in Machining Center

1) Vernier caliper

Vernier caliper is composed of six parts, main scale, vernier scale, internal measuring jaws, external measuring jaws, depth rod and lock screw.

① Main scale: the main scale is used to read the whole numbers that lines up with the vernier zero scale.

② Vernier scale: the vernier scale is used to take the numbers on the vernier scale that lines up perfectly with any line on the main scale.

③ Internal measuring jaws: the internal jaws are used to measure inside diameter.

④ External measuring jaws: the external jaws are used to measure outside diameter.

⑤ Depth rod: the depth rod is used to measure depth.

⑥ Lock screw: the lock screw is used to fix the vernier scale.

Before using the vernier caliper, clean the measuring jaws with a soft cloth and close the measuring jaws to make sure no light can be seen through them. Then, check whether the zero scale lines of the vernier scale and the main scale are aligned. The measurement can be performed if they are aligned. Otherwise, the zero error must be recorded. If the zero line on vernier scale is to the right side of that on the main scale, it is called positive zero error; if to the left side, it is called the negative zero error. This rule is consistent with that of the number axis — the reading to the right of the origin point is positive, and it is negative if to the left of the origin point.

In measuring, hold the scale with your right hand, move the vernier scale with your thumb, and hold the object to be measured in your left hand, making sure that the object is located between the measuring jaws. When the measuring jaws are against the object, the readings can be taken, the picture of vernier caliper is shown as in Figure 1-4-5.

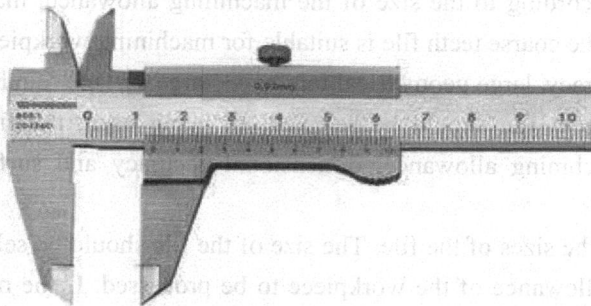

Figure 1-4-5　Vernier Caliper

In measuring the external dimensions of a part with the vernier caliper, the connection lines of the two measuring surfaces of the vernier caliper should be perpendicular to the measured surface and not be skewed. During the measurement, the vernier caliper should be slightly shaken and placed in the vertical position. Otherwise, the measurement result would be larger than the actual size. When using the vernier caliper, firstly, open the moveable jaw of the caliper so that the workpiece can easily fit into the space between the jaws, and attach the part to the fixed jaw; then slide the moveable jaw against the part with slight pressure. If the vernier caliper is equipped with an inching device, the lock screw on the device can be tightened at this time, and then adjust the thumb screw for the jaws to grip the part and take the readings. Do not adjust the two measuring jaws to be close to or even smaller than the measured size, and do not force the jaws against the parts. In this way, the measuring jaws are in deformation, or the surface to be measured is worn prematurely, causing the caliper to lose its due accuracy.

2) Dial caliper

Unlike the ordinary caliper, the dial caliper has a groove on the front of the main scale, and a rack is installed above the groove. A dial indicator with a graduation value of 0.01 mm or 0.02 mm is installed above the scale frame. A roller is installed at the tail of the ruler frame, and it rolls on the edge of the main scale surface to move the scale frame forward and backward, the picture of dial caliper is shown as in Figure 1-4-6.

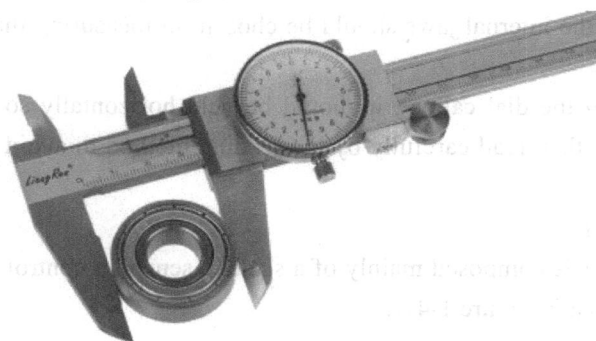

Figure 1-4-6   Dial Caliper

When using the dial caliper, follow the steps below.

① Before using, clean the vernier scale and then pull the scale frame to make sure the sliding can be flexible and smooth, and the scale should not be too loose or tight or even stuck. After the scale being fixed by lock screw, its readings should not change.

② Check the zero position. Gently close the caliper all the way to make sure no light can be seen through the jaws, and the dial indication needle is set to zero. Meanwhile, check whether the scale and the frame are aligned at the zero scale line.

③ In measuring, slowly push or pull the scale frame by hand to make the measuring jaws slightly contact on the surface of the part; then gently shake the dial caliper to keep the measuring jaws in good contact. In using the dial caliper, the operator has to control it totally relying on one's sense for there is no force-measuring mechanism. In order to make sure the measurement accuracy, excessive force is not allowed.

④ In measuring the external dimensions, the movable measuring jaw of dial caliper should be opened first so that the workpiece can be freely placed between the two measuring jaws; then the fixed measuring jaw is placed against one end of the workpiece and slide the scale frame by hand so that the movable jaw is closely attached to the other end of the workpiece. It should be noted that the two ends of the workpiece and the measuring jaws must not be inclined during measurement; the distance between the two measuring jaws should not be less than the size of the workpiece, otherwise, in such case it means that the part is forcibly tucked between the measuring jaws.

⑤ In measuring the internal diameter, the internal measuring jaws should be separated and the distance between the jaws should be smaller than the size to be measured. After the measuring jaws being placed into the hole, move the internal measuring jaws till them to be closely attached to the inner surface of the workpiece. Then, the readings can be taken. Please note that the measuring jaws of the caliper should be placed at the two ends of the diametrical line of the hole, and that the jaws must not be inclined.

⑥ Surfaces of different shapes can be measured by using the measuring jaws of dial caliper, and the jaws should be selected properly in terms of the shapes of parts to be measured. In measuring the length and external dimensions, the external jaws should be used; in measuring the internal diameter, the internal jaws should be chosen; in measuring the depth, the depth rod should be selected.

⑦ When reading the dial caliper, it should be held horizontally so that the operator can face the scale line and then read carefully by the reading method to avoid reading errors caused by carelessness.

3) Digital caliper

The digital caliper is composed mainly of a scale, a sensor, a control calculation part and a digital display, as shown in Figure 1-4-7.

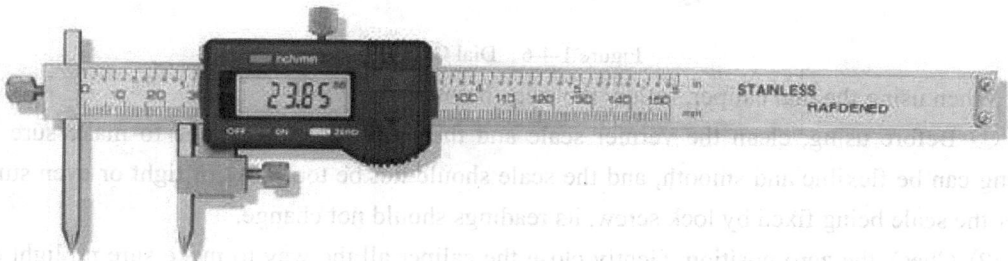

Figure 1-4-7   Digital Caliper

The working principle of digital caliper is to drive the circular grid plate to rotate through the high-precision rack mounted on the caliper body. Based on the photoelectric pulse counting principle, the displacement of caliper measuring claw is converted into pulse signal, and the measured size is displayed on the screen in the form of numbers through the counter.

In using the digital caliper, follow the steps below:

① Before using it, loosen the lock screw on the top of the meter, move the meter away smoothly, and clean the measuring surface and the guide surface with a cloth.

② Check whether each button is flexible and effective, and whether the digital display at any position is stable and clear before use.

③ Before measuring, the two external measuring surfaces of the digital caliper must be kept in contact. Press the ON / OFF button (turn on the meter), and then press the ZERO button to reset the LCD(Liquid Crystal Display) display to zero.

④ In measuring parts, read the measured value directly on the LCD display window.

4) Outside micrometer

The outside micrometer consists of a frame, an anvil, a spindle, a lock lever, a thread sleeve, a sleeve, a thimble, a nut, a joint, a force measuring device, a spring, a ratchet stop, a ratchet, etc., as shown in Figure 1-4-8.

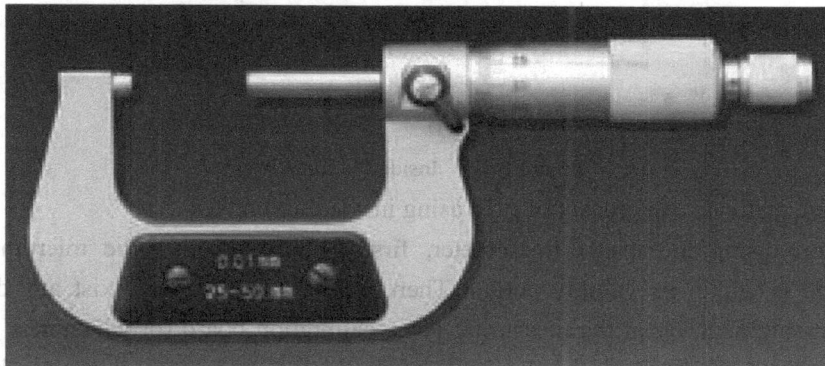

Figure 1-4-8　Outside Micrometer

In using the outside micrometer, follow the steps below.

① Before using the micrometer, firstly calibrate it to be at the zero position. Loosen the locking lever to clean oil stains, especially those on the contact surface of the anvil and the spindle. Check whether the end face of the thimble coincides with the zero scale line on the sleeve. If not, revolve the knob until the spindle is close to the anvil, and then rotate the force measuring device. When the spindle just touches the anvil, we hear a click and then stop rotating. If the two zero lines still do not match (we cannot be sure that the 0 on the thimble coincides the scale on the sleeve and the 0 of the movable scale coincides with the horizontal line of the sleeve), loosen the small screw on the sleeve and adjust the position of the sleeve with a special wrench to align the two zero lines, and then tighten the small screw. There are different methods for adjusting micrometers produced by different manufacturers. The above is only one of them.

To check the zero position of the micrometer, it is necessary to keep the spindle and anvil in a distance. However, sometimes the two surfaces might not be separated when the force measuring device is spun backward. At this time, the left side of the anvil on the frame can be pressed with the left palm and with the right palm pressed against the force measuring device, and then turn the knob counterclockwise with fingers to separate the spindle and the anvil.

② Read the whole millimeter and half millimeter on the sleeve first.

③ Then see which line on the thimble coincides with the center line of the sleeve, and take this reading as the decimal value less than half a millimeter. Finally, add up the two readings to get the measured size of the workpiece.

5) Inside micrometer

The inside micrometer is composed mainly of the micrometer unit, measuring jaws and various extension rods, as shown in Figure 1-4-9. The complete set of inside micrometer is equipped with adjustment tools for correcting the zero position of the micrometer unit.

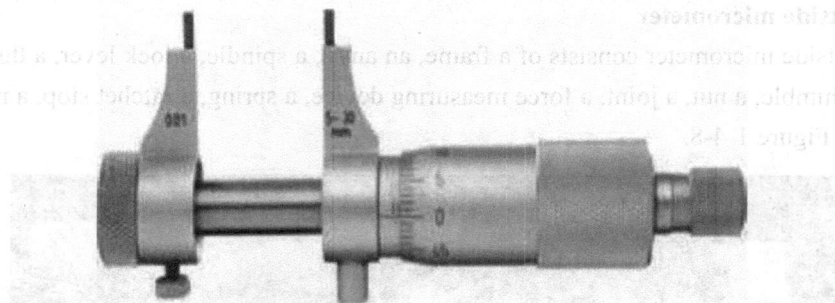

Figure 1-4-9   Inside Micrometer

Below are methods and precautions for using inside micrometer.

① Before using the inside micrometer, firstly check whether the micrometer has a certificate and is within its validity period. Then, check whether there exist any defects that affect the measurement, clean the measuring jaws, calibration board (or adjustment tools), and the measuring head. Then, check the flexibility of the thimble by rotating it, the condition of the spindle of the extension rod, and check whether the locking knob works reliably.

② Calibrate the measuring unit to be at the zero position with the calibration board with gentle and even force. If there is an error that affects the measurement accuracy, do not adjust it by yourself and send it to the measurement lab for testing in time.

③ Given below is the correct method for adding the extension rods onto the micrometer. If it is necessary to connect the extension rods, the minimum number of extension rods should be selected to form the required size and thus reduce the cumulative error. Meanwhile, the largest extension rod is connected with the measuring unit and the rest are connected with the measuring jaws in sequence to reduce the bending of the spindle. After the connection, repeatedly check whether the extension rods are fully tightened.

④ In measuring the diameter of a hole in a workpiece, the inside micrometer and the workpiece must be strictly isothermal. Avoid the heat from hands to be transferred to the micrometer. The following steps should be taken for measuring the diameter of a hole with an inside micrometer. Firstly, fit two ends of the measuring jaws against the area to be measured. Secondly, turn the thimble so that the measuring end (which is close to the thimble) swings in the radial section of the hole to find the maximum size and then swings in the axial section to find the maximum size. This adjustment needs to be repeated several times. Finally, tighten the locking screw and take out the inside micrometer and take the readings. In measuring the distance between two parallel planes, the micrometer should be swayed in multiple directions, and the minimum size should be taken as the measurement result.

6) Inner diameter gauge

The inner diameter gauge refers to a measuring instrument that changes the linear movement of the measuring jaws into the angular movement of the pointer and which consists of a measuring frame of the internal measuring lever type and a dial indicator. It is used to measure or inspect the inner hole diameter, deep hole diameter and shape accuracy of parts, as shown in Figure 1-4-10.

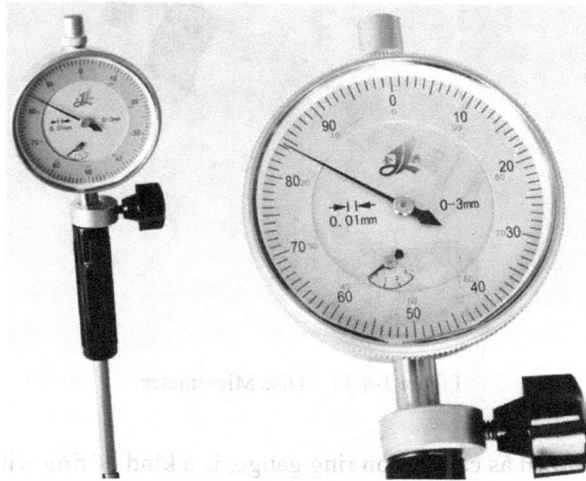

Figure 1-4-10　Inner Diameter Gauge

In using the inner diameter gauge, follow the steps below.

① According to the tolerance of the dimension to be measured, select a micrometer (the normal division value is 0.01).

② Adjust the dimension of micrometer to the nominal division value to be measured and tighten up the screw in the micrometer.

③ Hold the inner diameter gauge in one hand and the micrometer in the other, and place the measuring head of the inner diameter gauge between the two jaws of the micrometer for calibration. Make sure that the measuring rod of the diameter gauge is perpendicular to the micrometer as much as possible.

④ Adjust the dial indicator so that the pressure value on the gauge is about 0.2-0.3 mm, and set the gauge needle to zero. Adjust the error indication paddle on the bezel according to the tolerance of the dimension to be measured.

7) Disc micrometer

The disc micrometer is shown in Figure 1-4-11. The steps for its use are as follows.

① Before using it, check whether the micrometer is in good condition and whether there is an inspection certification.

② Then, clean the measuring faces of the micrometer and glass surface to be measured.

③ When using it, it must be reset to zero point first. Slowly turn the micrometer to make sure the disc-shaped spindle is in contact with the anvil. The force needed for the contact should be consistent with that for measurement. Considering the uncertainty of measurement, it is generally necessary to reset to zero point more than twice and the number of measurement should be no less than three times.

④ Place the part to be measured between two measuring surfaces, adjust the differential cylinder, once the working surface touches the measured object, adjust the force measuring device, stop when you hear three "clicks", and then you can read.

Figure 1-4-11   Disc Micrometer

8) Ring gauge

Ring gauge, also known as calibration ring gauge, is a kind of ring with specific dimensions and attributes used to check and calibrate other measuring tools, as shown in Figure 1-4-12.

Figure 1-4-12   Ring Gauges

As for the appearance, ring gauges can be divided into smooth ring gauge (also ring gauge), thread ring gauge, parallel thread ring gauge, etc.; as for different purposes and manufacturing standards, there are standard ring gauge, SK standard ring gauge, German standard ring gauge, pipe ring gauge, etc.; as for different materials, they can be divided into ceramic ring gauge and metal ring gauge.

9) Thread ring gauge

The thread ring gauge is used to measure the accuracy of the external thread, and to check and determine whether the thread size of the workpiece is qualified. There are two types of thread ring gauges: Go gauge one and No-Go gauge one.

(1) Go gauge.

① The Go gauge should be tested and measured by relevant inspection and measurement agencies before it can be put into use on the production site.

② As for its use, it should be noted that the tolerance grade and deviation code of the thread to be measured are the same as that marked on the gauge (for example, M24×1.5-6h and M24 × 1.5-5g, the two ring gauges have the same shape but of different thread tolerance grade. Improper use will lead to batches of non-conforming products).

③ For the inspection and measurement, first clean up the oil and impurities on the thread, then rotate the ring gauge with your thumb and index finger after it is aligned with and fit in the thread in free state.

The gauge must be perfectly fitted in the entire thread. Otherwise, it is judged as unqualified.

(2) No-Go gauge.

① The No-Go gauge should be tested and measured by relevant inspection and measurement agencies before it can be put into use on the production site.

② In using it, it should be noted that the tolerance grade and deviation code on the thread to be measured are the same as that marked on the gauge.

③ For the inspection and measurement, first clean up the oil and impurities on the thread to be measured; then rotate the ring gauge with your thumb and index finger after it is aligned with the thread. If the length of screw thread is within 2 pitches, it is qualified; otherwise, it is unqualified.

10) Plug gauge

Plug gauge, aka thread plug gauge, is a measuring tool. There are two commonly used ones, the round hole plug gauge and the thread plug gauge. The round hole plug gauge can be made from the maximum to the minimum solid sizes, as shown in Figure 1-4-13. The thread plug gauge is used for measuring whether the internal thread size meet the standard, as shown in Figure 1-4-14. According to the thread type, the thread plug gauge can be divided into three types, the one with normal coarse threads, with fine threads and with pipe threads. Thread plug gauges with a pitch of 0.35 mm or less in Class 2 accuracy and higher accuracy, or with a pitch of 0.8 mm or less in Class 3 accuracy have no end faces for measurement.

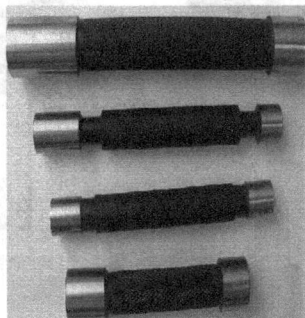

Figure 1-4-13   Round Hole Plug Gauges

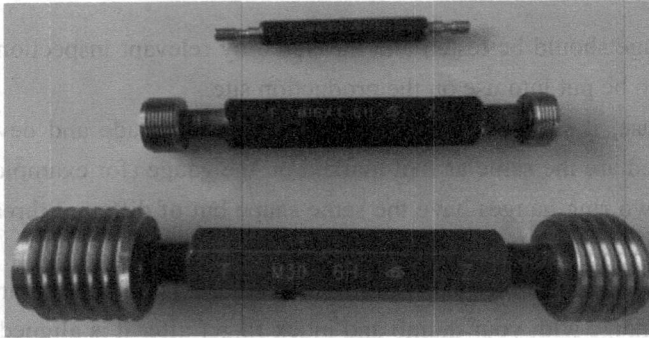

Figure 1-4-14   Thread Plug Gauges

The thread ring gauge or plug gauge is of multiple parameters. There are two methods for its inspection, the comprehensive inspection and single inspection. The former way uses the gauge to test the comprehensive results of geometric parameter deviation that affects the interchangeability of thread. These inspections include using plug gauge or No-Go gauge to check the effective diameter (including bottom diameter) and the single diameter of the thread to be measured, and using a smooth limit gauge to test the actual top diameter of thread. The thread plug gauge simulates the largest solid tooth of the thread to be measured, checks whether the effective diameter of the thread exceeds the diameter of the largest solid tooth, and simultaneously checks whether the actual size of the bottom diameter exceeds its maximum solid gauge.

11) Height gauge

The height gauge is also known as the height vernier caliper, as shown in Figure 1-4-15. Its main purpose is to measure the height of the workpiece. In addition, it is often used to measure the tolerance dimensions of shape and position, and sometimes to draw lines.

Figure 1-4-15   Height Gauge

In terms of the different ways to take the reading, the height gauges can be classified as the normal one and the digital one. The mechanical height gauge with counter is a different type.

The commonly used specifications of the height gauge are 0–300 mm, 0–500 mm, 0–1000 mm, 0–1500 mm, and 0–2000 mm.

In terms of the structure, there are single-column gauge and double-column gauge. Those with specifications of 0–300 mm and 0–500 mm are the single-column type.

The double-column type is used mainly for more precise or larger-scale measurement.

There are many types of double-column digital height gauges, including digital height gauge, dial digital height gauge and vernier digital height gauge.

In using the height gauge, follow the steps below.

① Before its use, wipe the surface of the protective sticker with a dry and clean cloth (or dipped in a little cleaning oil).

② Working environment: keep the temperature between 5–40℃; the relative humidity is 80%; prevent any hydrous liquid from contacting the protective sticker.

③ Do not apply a voltage to any part of the gauge (such as engrave words with an electric pen), lest a damage be done to the circuit.

④ Be careful with the sharp tip of the measuring jaw.

12) Depth gauge

The depth gauge is composed of a main scale and an auxiliary scale (or vernier). There is a locking screw on the top of the auxiliary scale to keep the reading fix, as shown in Figure 1-4-16. The minimum unit of the main scale is 1 mm; the vernier scale is 49 mm in length divided into 50 equal parts. Depth gauge is the most basic measuring tool in the mechanical manufacturing and maintenance industry, which is used to measure the length and depth of components.

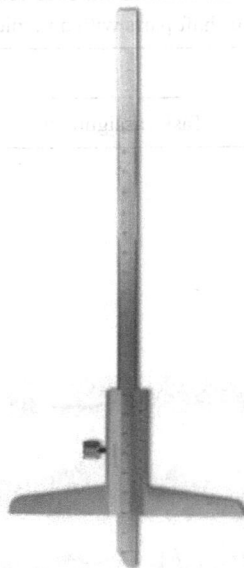

Figure 1-4-16　Depth Gauge

For different needs, you can choose different measurement ranges for depth gauges (or different length of main scales), such as 0–200 mm, 0–300 mm, 0–500 mm, 0–1000 mm, etc.

In taking the reading of the depth gauge, firstly read the main scale. Take the value of 1 mm at the main scale to align with the left of zero at vernier scale (aka auxiliary scale). Then, read the vernier scale. Observe alignment between two scales and take the value of vernier scale. At last, sum the two values to get the size of the workpiece being measured.

The specific reading method is shown in Figure 1-4-17. When the 0 scale line at the left end on the vernier scale is aligned with the 1 mm scale line on the main scale, take the reading from the main scale, which is 1.00 mm. Slightly move the vernier scale to the right so that the next scale line align with the 0 scale line on the left end of the vernier scale is aligned with a scale line of the main scale, and then the reading is $1 + 1 \times 0.02$ mm $= 1.02$ mm and keep moving the vernier scale like this, and the readings are 1.04 mm, 1.06 mm, …, 1.98 mm respectively. When the 0 at the left end on the vernier scale is aligned with a certain scale line on the main scale (such as 2 mm on the main scale), directly take the reading from the main scale, which is 2 mm. Therefore, the measurement accuracy of the depth gauge is 0.02 mm.

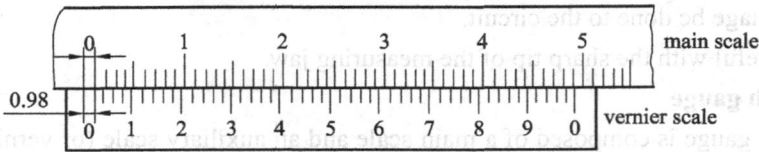

Figure 1-4-17   Readings on Main Scale and Vernier Scale

3. Skills Training

Complete the task sheet as Table 1-4-1 based on what you have learned above.

Table 1-4-1   Task Sheet

Task name	Measure shaft parts with a vernier caliper and micrometer		
Class		Group No.	
Tasks assignment			
Parts to be measured			

Continued Table

Measurements	

Task 5　Basic Mold Fixtures and its Usage Methods

【Knowledge objectives】

(1) Be familiar with basic mold fixtures in machining center.

(2) Be familiar with workpiece clamping on the vise.

(3) Be familiar with workpiece clamping with a clamping plate.

【Ability objectives】

(1) Master the use of basic mold fixtures in machining center.

(2) Master the method for clamping a workpiece on the vise.

(3) Master the method for clamping a workpiece using a clamping plate.

1. Mold Fixtures in Machine Center

1) Flat vise

Flat vise is one of the main fixtures for planer, milling machine, drilling machine, grinders, and slotting machine. It is widely used for processing various planes, grooves, angles, etc. on milling machines, drilling machines, etc.

Flat vise is a kind of general-purpose accessory for machine tools. Together with the worktable, they are used to fix, clamp and position the workpiece during processing, here we call them as machine flat vise. The machine flat vise is composed of base, moveable jaw, nut, handle and other components. According to its structure and purpose, machine flat vise can be divided into the following types, general flat vise, flat vise for angle pressing machine use, tiltable machine flat vise, ultra-precision machine flat vise for machine use, flat vise for force increasing machine use, etc. Figure 1-5-1 shows the general flat vise.

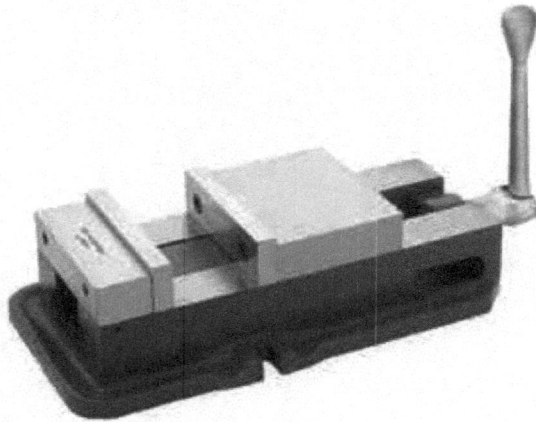

Figure 1-5-1　A general Machine Flat Vise

2) Hydraulic vise

The hydraulic vise as shown in Figure 1-5-2 is an improvement on the existing screw-driven bench vise, which is used mainly for mass production. It can achieve quick clamping and loosening, and it can guarantee the clamping force. The use of hydraulic vise can avoid the debugging process due to the fact that the clamping force cannot be determined when clamping a thin part. At the same time, its use can achieve fast clamping and loosening, thus greatly improving the production efficiency. In order to achieve its fast clamping and loosening, the traditional thread transmission is changed into hydraulic transmission, and the movable vise is controlled by hydraulic cylinder, so as to realize the fast movement of the movable vise, while the clamping force is guaranteed by the overflow valve in the hydraulic system. We can adjust the pressure of the overflow valve to ensure the clamping force. The hydraulic vise has the following functions.

① Workpiece clamping — tightening and loosening the workpieces.

② Motion — reciprocating linear motion.

③ Mechanism — to obtain reciprocating linear motion, including screw transmission mechanism, crank slider mechanism, pneumatic transmission mechanism and hydraulic transmission mechanism.

Figure 1-5-2   Hydraulic Vise

The hydraulic vise performs as the following.

① Cycle period — no more than 5 seconds.

② Motion accuracy — average.

③ Work efficiency — high.

④ Reliability — 5000 hours service life.

3) Pneumatic vice

Pneumatic vise, also known as pneumatic flat vise, pneumatic bench vise, etc., refers to a kind of fast clamping tool suitable for mass production on double-axis compound machine, CNC machining center, milling machine, drilling machine, etc., its picture is shown as in Figure 1-5-3. It is characterized by its being time-saving, labor-saving, and high in efficiency, greatly reducing the labor intensity in the production process and improving production efficiency. According to the working condition, you need to produce fixture applicable to certain product and install it on the jaw.

Figure 1-5-3   Pneumatic Vice

The pneumatic vise is composed mainly of air cylinder, vise base, jaw slider, guide shaft, adjusting nut, etc. For it to be firmly fixed on the workbench, the two screws for fixing must be tightened. If it can not be attached securely when processing, the pneumatic vise may be damaged so as to affect the following processing.

2. Workpiece Clamped on the Clamping Tool of Flat Vise

1) Aligning in the flat vise (in groups)

The flat vise must be calibrated before use. Accuracy calibrating for flat vise is to ensure the relative position accuracy of the workpiece being processed. The method for calibrating the flat vise is as follows.

① Use a scribing block or a pin to roughly align the accuracy of flat vise. Firstly, apply some lubricant at the back of the pin and stick it to the cutter head; secondly, press the fixed jaw of the vise towards the tip of the pin to make sure the pin or scriber is about 1 mm away from the fixed jaw; then slowly shake the longitudinal workbench and observe whether the distance from the tip of the pin to the surface of fixed jaw always keeps at 1 mm. If not, loosen the fastening screws on both sides of the flat vise for adjustment until the distance generally keeps at 1 mm.

② Use the dial indicator to accurately align the accuracy of flat vise. In aligning the flat vise, put the magnetic base of the dial indicator on the surface of the beam guide rail or on the spindle part of the vertical milling head, so that the measuring rod is perpendicular to the plane of the fixed jaw and the measuring head is in contact with the plane of the jaw. Then, compress the measuring rod for about 0.3~0.5 mm, move the workbench longitudinally, and observe the readings of the dial indicator. If we want to keep the readings the same, the fixed jaw must be parallel to the feed direction of the workbench, thus, good position accuracy can be obtained in processing.

③ After the aligning of the fixed jaws staying parallel to the feeding direction of the workbench, align the verticality of the fixed jaws and the plane of the workbench in the same way.

2) Method for clamping the workpiece on flat vise

① Workblank clamping. When clamping the workblank, a flat workblank surface should be selected as the rough reference surface and it should be close to the fixed jaw of the vise. In clamping, copper sheet should be placed between iron jaw plane and workblank.

② Machined surface clamping. When clamping the workpieces that have been roughly machined, one machined surface should be selected as the reference surface, and it should be close to the fixed jaw or the guide rail of the vise. When the machined surface is close to the fixed jaw, a round bar can be placed between the movable jaw and the workpiece, so as to ensure that the reference surface fits well into the fixed jaw.

3. Use of the Clamping Plate to Clamp the Workpiece

1) Clamping plate

For medium-sized, large-sized and complex-shaped parts, clamping plate is generally used to fasten the workpieces on the worktable of CNC milling machine. The tools used for clamping workpieces with the clamping plate are relatively simple, including the plate, iron pads, T-bolt and T-nut, bolt and nut, as shown in Figure 1-5-4. However, in order to meet the requirements for clamping parts of different shapes, there are many types of clamping plates. For example, the box parts are usually installed on the workbench by three-sided installation or by one plane and two pin holes for positioning and then fixed by clamping plate.

Figure 1-5-4   Clamping Plate

2) The process of workpiece clamping

Below is the process of workpiece clamping.

① Fix the positioning pin in the T-shaped groove of the machine tool and place the shim plate on the workbench.

② Select the appropriate clamping plate, step-shaped block and T-bolt, place them in corresponding positions.

③ Clamp the part.

3) Precautions for clamping with a clamping plate

Take the following precautions for clamping with a clamping plate.

① The milling part of the workpiece must not be pressed by the clamping plate, so as not to hinder the normal milling process.

② The height of iron pad below clamping plate should be appropriate, so as to prevent the poor contact between the plate and part.

③ When clamping thin-walled workpieces, the clamping force should be appropriate.

④ The bolt should be as close to the workpiece as possible to increase the clamping force.

⑤ The copper pad must be placed between the smooth surface of the workpiece and the clamping plate to avoid damaging the surface of the workpiece.

⑥ The part of the workpiece being pressed should not be suspended. If there is suspension, it should be padded.

⑦ When the rough workpiece is directly clamped on the worktable of the milling machine, a paper pad or a copper pad should be applied between the workpiece and the worktable's surface. In this way, not only the worktable can be protected, but also the friction between the worktable and the workpieces can be increased for clamping safety.

4. Skills Training

Complete the task sheet as Table 1-5-1 based on what you have learned above.

Table 1-5-1　Task Sheet

Task name	Alignment of the flat vise		
Class		Group NO.	
Tasks assignment			
Flat vise			
Procedures for its alignment			

Task 6　Installation and Adjustment of Basic Cutting Tools in Machining Center

【Knowledge objectives】

　　(1) Understand the structure of the basic cutters of machining center.

　　(2) Learn how to use the cutter holder of machining center.

【Ability objectives】

　　(1) Master the method for selecting cutters for different workpieces.

　　(2) Master the method for using the cutter holder.

1. Structure of Basis Cutters of Machining Center

1) Milling cutters and milling cutting tools

Cutters, as a main part in cutting tools, plays an important role in cutting work. The HSS cutters in milling cutting tools are shown in Figure 1-6-1.

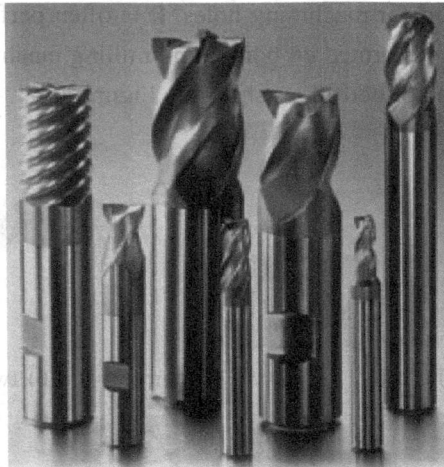

Figure 1-6-1　HSS Cutters with no Coating

Typical milling cutters (cutting tools) in machining center are shown in Figure 1-6-2.

　　① End milling cutting tool(also known as rotary blade): it is used to remove more materials.

　　② Cylindrical end milling cutting tool: it is used to process the right-angled profile with vertical flanges.

　　③ Insert-type shaft milling cutting tool: a multi-head cutting tool that arranges the cutters spirally to produce a special "smooth" machining effect.

　　④ Long hole milling cutting tool (also known as groove milling cutting tool): it is used to cut the center surface, and to be inserted into deep grooves for cutting. It usually has 2 to 3 cutting edges.

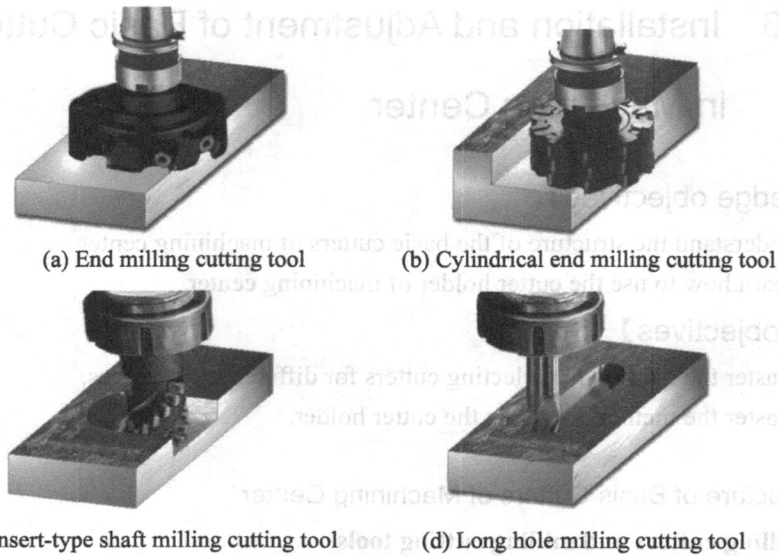

(a) End milling cutting tool　　　　(b) Cylindrical end milling cutting tool

(c) Insert-type shaft milling cutting tool　　(d) Long hole milling cutting tool

Figure 1-6-2　Milling Cutters

2) Drill

Drilling is a basic method for machining holes. It is often performed on drilling machines and lathes, and it can also be performed on boring and milling machines. Drill mainly include U drill, twist drill, center drill and pilot drill, as shown in Figure 1-6-3.

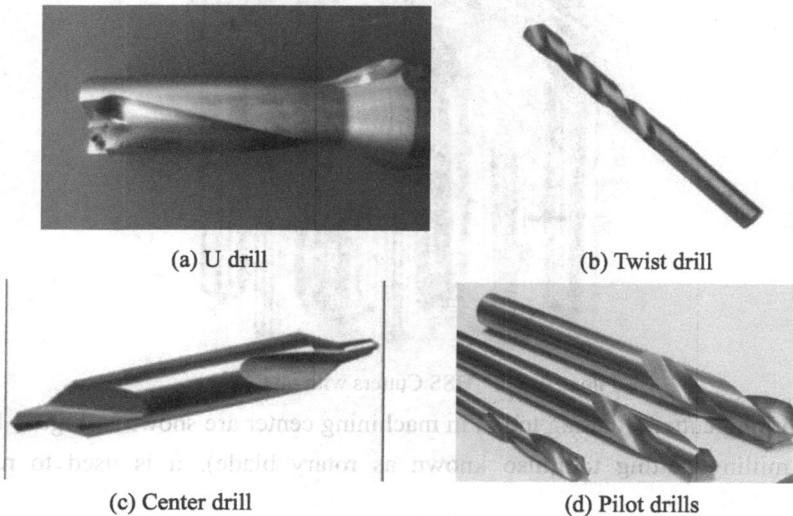

(a) U drill　　　　　　(b) Twist drill

(c) Center drill　　　　　　(d) Pilot drills

Figure 1-6-3　Drill

3) Reamer

Reamer is a rotary finishing tool with a straight or spiral edge. It is used for reaming or repairing holes. Because of small cutting value, its machining accuracy is usually higher than that of the drill. It can be manually operated or installed on the drilling machine, picture of reamers is shown as in Figure 1-6-4.

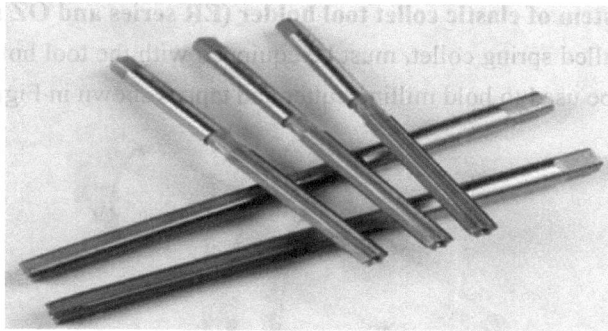

Figure 1-6-4   Reamers

4) Tap

Tap is a tool for processing internal screw threads. In terms of its shape, tap can be divided into spiral tap and straight-edge tap; in terms of its use environment, tap can be divided into hand tap and machine tap; in terms of its specifications, tap can be divided into metric tap American tap, and English tap; in terms of its origin, tap can be divided into imported tap and domestic tap. Tap is the most important tool for manufacturing operators to process screw threads, picture of taps is shown as in Figure 1-6-5.

Figure 1-6-5   Taps

5) Boring cutting tool

Boring cutting tool is used for finishing holes. A special attention should be paid in cleaning the installed boring cutting tool. The cutter in it should be pre-adjusted before machining, and the dimensional accuracy, intact condition must meet the requirements, and then we process for trial boring conducted to determine the size. Boring cutting tools are shown in Figure 1-6-6.

Figure 1-6-6   Boring Cutting Tools

2. Tool Holder in Machining Center and Its Use

1) Clamping system of elastic collet tool holder (ER series and OZ series)

ER collet, also called spring collet, must be equipped with the tool holder of ER collet. ER collet tool holder can be used to hold milling cutter and tap, as shown in Figure 1-6-7.

Figure 1-6-7　ER Collet tool holder

ER collet is of strong clamping force, wide clamping range and good precision. It is widely used in processing such as boring, milling, drilling, tapping, grinding and engraving. The ER collet tool holder is generally used with the CNC tool holder. At present, the main standards include BT, SK, CAPTO, BBT, HSK and other specifications of spindle models.

2) Powerful tool holder

Powerful tool holder that is of high-speed, high-precision, high-efficiency processing fixture, requires that the tool holder should be able to clamp the tool with high precision on the machine tool, and still maintain a high-precision clamping state during the process, picture of the powerful tool holder is shown as in Figure 1-6-8.

Figure 1-6-8　Powerful Tool Holder

Below are the advantages of powerful tool holder.

① High in system accuracy. It includes system positioning clamping accuracy and tool repeat positioning accuracy, the former refers to the connection accuracy between the powerful tool holder and the CNC tool holder, and between the powerful tool holder and the spindle of machine tool. The latter refers to the consistency of system accuracy of tool holder after each tool change. The tool system is high in system accuracy, which can guarantee the static and dynamic stability of the tool system under conditions of high-speed machining.

② High in system rigidity. The static and dynamic rigidity of the tool system is an important factor that affects the machining accuracy and cutting performance. Insufficient rigidity of the tool system leads to the vibration of the tool system, thereby reducing machining accuracy, aggravating the wear of the tool and reducing the service life of the tool.

③ Good in dynamic balance. Under high-speed cutting conditions, the imbalance of tiny masses causes huge centrifugal force, which leads to the sharp vibration of machine tool during the processing. Therefore, the dynamic balance of the high-speed tool system is very important.

3. Skills Training

Complete the task sheet as Table 1-6-1 based on what you have been learned above.

Table 1-6-1   Task sheet

Task name	Clamping end mill of 8 mm diameter		
Class		Group No.	
Tasks assignment			
Steps for clamping			
Precautions			

Module 2

Basic Programming in Machining Center

Task 1   Geometric Coordinate System of the Machine Tool

in Machining Center

【Knowledge objectives】

(1) Understand the coordinate system of CNC machine tool.

(2) Understand points in the working area of machine tool in machining center.

(3) Learn to determine the programming coordinate system and calculate the base point.

【Ability Objectives】

(1) Master the methods for identifying the coordinate system of machine tool.

(2) Master the methods for calculating the base point.

1. Knowledge of the Coordinate System of CNC Machine Tool

The coordinate axes and movement directions of CNC machine tool are specified in order to accurately describe the movement of the machine tool, simplify the programming, and make the programs interchangeable. International Organization for Standardization has now unified the standard coordinate system, and China has also issued the corresponding standard (JB305-82), which explicitly stipulates the coordinates and movement directions of CNC machine tools.

1) The principle for naming the movement direction

In order to determine the machining process of the machine tool in the drawing, there is specification that the tool is always assumed to move relative to the coordinate system of the stationary workpiece.

2) The provisions for coordinate system

In order to determine the movement direction and moving distance of the machine tool, a coordinate system should be established on the machine tool, which is the standard coordinate system. In programming, the standard coordinate system is used to define the direction and distance of movement.

The coordinate system on the CNC machine tool follows the right-handed Cartesian coordinate system. In Figure 2-1-1, the direction of the thumb is the positive direction of X axis, the index finger points to the positive direction of Y axis, and the middle finger points to the positive direction of Z axis.

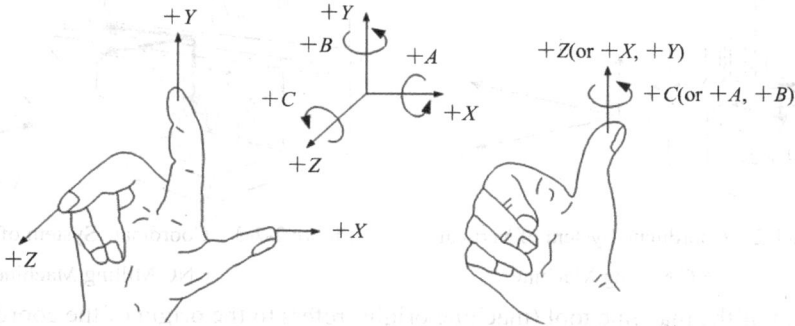

Figure 2-1-1   Right-handed Cartesian Coordinate System

Below are some common coordinate systems.

(1) The coordinate system of machine tool. The coordinate system of machine tool is an inherent part of the machine tool, and its orientation is determined by referring to some benchmarks on the machine tool. There are some fixed reference lines on the machine tool such as the centerline of the spindle, and some fixed reference surfaces such as the worktable, the end surface of the spindle, the side surface of the worktable, the guide rail surface, etc. Different machine tools have different coordinate systems. In coordinate standard, the coordinate axis of the tool movement parallel to the spindle (transmitting cutting force) of the machine tool is specified as Z axis, and the direction in which the tool is away from the workpiece is the positive direction of Z axis (+Z). If the machine tool has multiple spindles, specify the one that is perpendicular to the clamping surface of the workpiece as Z axis and specify X axis to be horizontal, perpendicular to Z axis and parallel to the clamping surface of the workpiece. For machine tools (lathes, grinders) whose workpiece rotates, specify the tool movement direction parallel to the horizontal slide (radial direction of workpiece) as X axis coordinate, at the same time, take the direction of the tool away from the workpiece as the positive direction of X axis. For machine tools (such as milling machines and boring machines) whose tool rotates, if Z axis is horizontal, look along the rear end of the tool spindle toward the workpiece, and the right direction is the positive direction of X axis; if Z axis is vertical, look from the spindle to the column, the positive direction of X axis points to the right for a single-column machine. For a double-column machine, look from the spindle to the left column, the positive direction of X axis points to the right. All the above positive directions are in terms of tool movement relative to the workpiece.

If the positive directions of X axis and Z axis having been determined, the positive direction of the Y axis can be determined according to the right-handed Cartesian coordinate system, that is, in the Z-X plane, when turning from $+Z$ to $+X$, the right spiral should move along $+Y$ direction. The common coordinate directions of machine tools are shown in Figure 2-1-2 and Figure 2-1-3. The directions shown in the figures are the moving directions of actual moving parts.

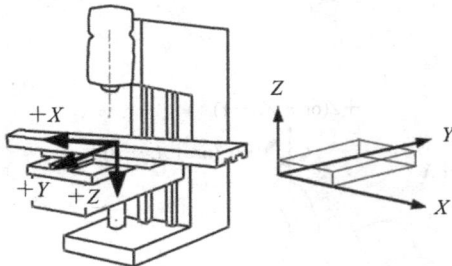

Figure 2-1-2   Coordinate System of Vertical CNC Milling Machine

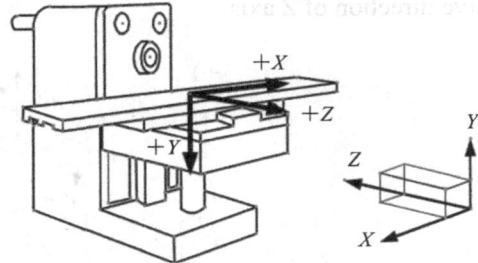

Figure 2-2-3   Coordinate System of Horizontal CNC Milling Machine

The origin of the machine tool (machine origin) refers to the origin of the coordinate system of the machine tool, and its position is usually at the maximum limit of each coordinate axis.

(2) The coordinate system of the workpiece. The coordinate system of the workpiece is used by programmers for programming and processing, and is the reference coordinate system of the program. The position of the coordinate system of the workpiece is based on the coordinate system of the machine tool. Generally, six coordinate systems of the workpiece can be set in a machine tool. The programmer takes a certain point in the drawing of the workpiece as the origin of the coordinate systems of the workpiece, which is called the working origin. The coordinate points of tool path during programming are determined according to the coordinates of the workpiece contour in its coordinate systems. In processing, the workpiece is installed on the machine tool with the fixture. At this time, the distance of the working origin from the machine origin is measured. This distance is called the origin offset(OO') of the workpiece, as shown in Figure 2-1-4.

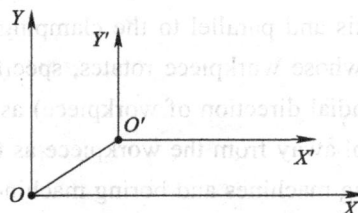

Figure 2-1-4   Coordinate System of Workpiece and Coordinate System of Machine Tool

This offset value must be stored in the CNC system before executing the machining program. During processing, the origin offset of the workpiece can be automatically applied to the coordinate system of the workpiece, so that the CNC system can determine the absolute coordinate value during processing according to the coordinate system of the machine tool. Therefore, programmers can use the origin offset function of the CNC system to compensate for the position error of the workpiece on the worktable without considering the actual installation position and installation accuracy of the workpiece on the machine tool. Most CNC machine tools now perform this function, which is very convenient to use.

(3) Additional movement of coordinate system. Generally we call X, Y and Z as the main coordinate system or the first coordinate system, if there are the second and third coordinate systems parallel to the first coordinate system, they are designated as U, V and W, and P, Q and R. The so-called first coordinate system refers to the linear movement direction close to the spindle. The one slightly farther away is the second coordinate system, and the third coordinate system is even farther away.

2. Points in the Working Area of the Machine Tool in Machining Center

1) Machine zero point

The machine zero point (also called the origin of the machine coordinate system) is the design origin of the machine coordinate system. Its position on the mechanical hardware, determined by the manufacturer, is a fixed point on the machine tool. It is a reference point not only for establishing the coordinate system of a workpiece on the machine tool, but also for debugging and processing of the machine tool. Now there are more and more types of machine tools, and the origins of machine tools designed by different manufacturers are also different. Usually, the origin of the lathe is set at the intersection of the end surface of the chuck and the center line of the spindle, while the origin of the milling machine is set at the limit position of the movement in the positive direction of the X, Y, and Z axes.

2) Machine reference point

The machine reference point is another point defined by the machine tool manufacturer. The coordinate position relationship between the machine reference point and the machine zero point is fixed. The position parameters of the reference point are stored in the CNC system. Returning to the reference point helps find the machine zero point. The machine reference point is generally located at the edge of the processing area of the machine tool. When the numerical control device is powered on, the machine zero point is unknown. In order to determine the machine coordinate system correctly when the machine is working, a machine reference point (starting point for measurement) is usually set within the movement range of each coordinate axis. When the machine tool is started up, the reference point is usually returned automatically or manually to establish the machine coordinate system and activate the parameters.

In CNC machine tools, a photoelectric encoder is usually used to return the machine to the reference point. The photoelectric encoder is installed at the end of the motor or the lead screw. In terms of the different structure of the photoelectric encoder, there are two ways to determine the machine reference point. The first is to determine it by using the relative position detection device. Because the position data of the relative position detection device (such as the relative photoelectric encoder) are lost after the machine is shut down, the machine has to return to zero point for processing operation every time after the machine is turned on. In processing operation, the block-type zero-point return is generally used. The second is to determine it by using the absolute position detection device. The absolute position detection device (such as the absolute photoelectric encoder) can detect the movement of the machine even after the power is cut off, so the machine tool does not have to return to the origin point every time after it is turned on. Since the position data are not lost after shutdown, it is high in reliability. If the absolute position detection device is replaced or the absolute position is lost, the reference point should be reset. The absolute position detection system generally uses zero-point return without block.

3. Determination of the Programming Coordinate System and Calculation of the Base Point

1) The programming coordinate system

The programming coordinate system, aka the workpiece coordinate system, refers to the coordinate system for programmer to calculate the coordinate point value. The zero point of the coordinate system of the workpiece is called as the programming zero point.

2) The base point

The intersection or tangent point of different geometric prime lines that constitute the contour of the part is called the base point. The base point can be directly used as the starting point or end point of its movement trajectory.

In determining the programming coordinate system and base point as in Figure 2-1-5, in the standard drawing, the lower left corner of the part is taken as the base point, on which the rectangular coordinate system is established.

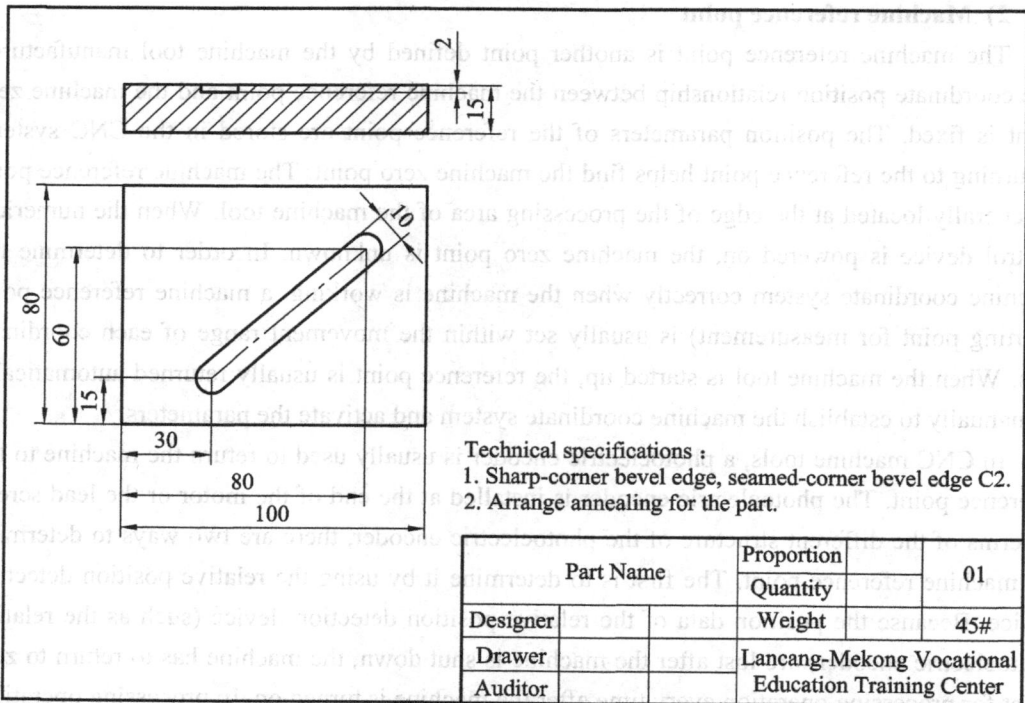

Technical specifications :
1. Sharp-corner bevel edge, seamed-corner bevel edge C2.
2. Arrange annealing for the part.

Part Name		Proportion		01
		Quantity		
Designer		Weight		45#
Drawer		Lancang-Mekong Vocational Education Training Center		
Auditor				

Figure 2-1-5　Programming Coordinate System and Base Point

4. Skills Training

Complete the task sheet as Table 2-1-1 based on what you have learned above.

Table 2-1-1　Task sheet

Task name	Establishment of the coordinate system of simple parts		
Class		Group No.	
Tasks			
Drawings of parts	Technical specifications : 1. Sharp-corner bevel edge, seamed-corner bevel edge C2. 2. Arrange annealing for the part.		
Procedures for establishing the coordinate system			

In the drawings table area:

Part Name		Proportion	01
		Quantity	
Designer		Weight	45#
Drawer		Lancang-Mekong Vocational Education Training Center	
Auditor			

Task 2　Basic Manual Programming in Machining Center

【Knowledge objectives】

(1) Understand the machining program and its basic format in machining center.

(2) Know the basic programming instructions in machining center.

(3) Learn the manual programming of plane milling.

【Ability objectives】

(1) Master the basic format of programming.

(2) Master the commonly-used basic programming instructions.

(3) Master the manual programming of plane milling.

1. The Machining Program and its Basic Format in Machining Center

The instructions settings sent to the CNC to run the machine in CNC machining center is called a program. According to the instructions, the tool moves along a straight line or an arc, and the spindle motor rotates or stops. In the program, instructions are set in the order along which the tool actually moves. A group of single-step sequential instructions is called a block. A block begins with the sequence number for identifying the block and ends with the end code of the block. In this book, " ; " or "Enter" is used to denote the end code of a block (EOB in ISO code and CR in EIA code). The machining program is composed of several blocks; and each block is composed of one or several instruction words, which represents a certain information unit; each instruction word is composed of address symbols and numbers, which represents a position or an action of the machine tool. There should be "EOB" or "CR" at the end of each block to indicate the end of the block and transferring to the next block. The address symbol is made up of character letter, number and symbol are called as character. An example of a block structure is shown in Table 2-2-1, and Table 2-2-2 is an explanation of the address symbols in the block example.

Table 2-2-1　Example of a block

Block content	Note
%	Start symbol
O0004	Block1 number
N1　G90G54G00X0Y0S1000M03;	First part block
N2　Z100.0;	Second part block
N3　G41 D01X20.0Y10.0;	⋮
N4　Z2.0;	
N5　G01Z-10.0F100;	
N6　Y50.0F200;	

Continued Table

N7 X50.0;	
N8 Y20.0;	
N9 X10.0;	
N10 G00Z100.0;	
N11 G40X0Y0M05;	
N12 M30;	Block finished

Table 2-2-2   Meanings of Common Address Symbols

Address symbol	Function	Meaning
A	Coordinate word	Rotate around X axis
B	Coordinate word	Rotate around Y axis
C	Coordinate word	Rotate around Z axis
D	Compensation number	Instruction of compensating for cutter radius
E	–	Second feed function
F	Feed rate	Feeding rate instruction
G	Get ready function	Instruction of conduct method
H	Compensation number	Compensation number assignment
I	Coordinate word	X axis direction coordinate at arc center
J	Coordinate word	Y axis direction coordinate at arc center
K	Coordinate word	Z axis direction coordinate at arc center
L	Loop time	Fixed circle and the loop times of sub-block
M	Auxiliary function	Machine open/close instruction
N	Number	Sequence number of block
O	Order number	Assignment of order number and sub-order number
P	–	Pause time or sequence number directing to certain function start at block
Q	–	Section number of fixed circle or fixed distance in circle
R	Coordinate word	Assignment of fixed distance or arc radius
S	Spindle speed function	Instruction of spindle rotation speed
T	Tool function	Instruction of cutter identifier number
U	Coordinate word	Increment coordinate value parallel to X axis
V	Coordinate word	Increment coordinate value parallel to Y axis
W	Coordinate word	Increment coordinate value parallel to Z axis
X	Coordinate word	Absolute coordinate of X axis or pause time
Y	Coordinate word	Absolute coordinate of Y axis
Z	Coordinate word	Absolute coordinate of Z axis

Block format refers to the order in which the command words are arranged in the block. Different CNC systems have different block formats. If the format does not meet the regulations, the CNC device sends an alarm and does not execute it. The common block formats are shown in Table 2-2-3.

Table 2-2-3　Common Block Format

1	2	3	4	5	6	7	8	9	10	11
N_	G_	X_ U_ Q_	Y_ V_ P_	Z_ W_ R_	I_J_K_ R_	F_	S_	T_	M_	LF
Number	Get ready function	Coordinate word				Feed function	Spindle speed function	Tool function	Auxiliary machine function	List finished

① Number of a block (number for short): it is usually represented by 4 digits, that is, "0000~9999", and the identification symbol "N" precedes number in format of N0001.

② Get ready function (G function for short): it consists of the address character "G" for preparation function and two digits.

③ Coordinate word: it consists of coordinate address symbols and numbers, and is arranged in a certain order. Each group of numbers must use a letter (such as X, Y, etc.) as the address code at the beginning. The address symbols for each coordinate axis are generally arranged in the following order,

X, Y, Z, U, V, W, Q, R, A, B, C, D, and E

④ Feed function F: it consists of feed address symbol F and numbers. The numbers, generally four-digit codes, indicate the selected feed speed, and its unit is generally "mm / min" or "mm / r".

⑤ Spindle speed function S: it consists of spindle address character S and numbers. The numbers indicate the spindle speed, and its unit is "r / min".

⑥ Tool function T: it consists of address character T and numbers, and is used to specify the tool number.

⑦ Auxiliary machine function (M function for short): it consists of auxiliary operation address character "M" and two digits. The codes for M function have been standardized.

⑧ List finished: it is listed after the last useful character in the block, which means the end of the block.

It should be noted that there are many standards for the command format of CNC machine tools worldwide, and that they are not completely consistent. With the development of CNC machine tools, the CNC systems have been continuously improved and innovated, and their functions are more powerful and convenient to use. However, there are certain differences in the program format among different CNC systems. Therefore, in operating a specific CNC machine tool, it is necessary to carefully understand the programming format of its CNC system.

2. Basic Programming Instructions in Machining Center

1) Auxiliary machine function (M function)

The auxiliary machine function, also known as M function, is the function to command the auxiliary action of the machine tool. Some M codes are defined by the system manufacturer, while others are defined by the machine tool manufacturer.

Generally, only one M code is allowed in a block, and up to three can be specified in some machine tools.

M00 — Stop for a program. After executing the program containing this command, the rotation, feeding and flow of cutting fluid of the spindle are all stopped, so as to carry out a certain manual operation, such as tool change, workpiece re-clamping, size measurement of the workpiece, etc. After restarting the machine tool, the remaining program is to be executed.

M01 — Planned stop (selective stop). M01 is basically similar to M00 in function. The only difference is that M01 is effective only if the selective stop key is pressed, otherwise, machine tool continues to execute remaining blocks. This instruction is generally used to check conditions such as the key dimensions. After checking, press the "Start" button to continue to execute the remaining program.

M02 — End of a program. This command is compiled in the last block. It means that after all commands in the program are executed, the spindle stops, feed stops, the flow of cutting fluid is turned off, the machine tool is in a reset state, and when the machine tool screen displays CRT, the program is finished.

M30 — End of the program and return to the head of the program to prepare for the processing of the next workpiece. When machine tool screen displays CRT, the program starts.

M06 — When the tool exchange between the spindle tool and the tool in the change position of the tool bank is executed, first, the machine implement the quasi-stop action of the spindle, then the tool change action is executed.

2) Spindle speed function (S function)

The rotation speed of the spindle is specified by the S code, and its unit is generally r / min. For example, S1200 indicates that the speed of the spindle is 1200 r / min. After S code is specified, that the spindle rotates or not, and forward or backward all depend on the M code.

3) Feed function (F function)

(1) Rapid feed speed. It is used to specify the moving speed in rapid traverse. The rapid feed speed is set by the parameters for each axis and does not need to be specified in the program. The rapid feed speed can be adjusted with the quick adjustment switch on the machine panel, which are of four gears, F0, 25%, 50%, 100%. F0 is the speed set by the parameters for each axis, and the rest are the override of the rapid traverse speed of each axis.

(2) Cutting feed speed. The cutting feed speed of the tool is specified by the value following the F code, that is, F_, which is used in linear interpolation, arc interpolation and fixed cycles. When moving in a straight line, F is the speed along the straight line. When moving in an arc, F is the speed in the tangential direction.

① Feed rate per minute (F94): the value following the F94 code is the feed rate per minute. In metric units, the linear speed unit of F is mm/min. In inch units, it is inch/min. If it is a rotary shaft, the unit of speed is (°) / min.

② Feed rate per spin (F95): the value following the F95 code is the feed rate per spin per minute. At this time, the spindle of the machine tool must be equipped with a position encoder.

③ Feed rate magnification: the feed rate per minute can be adjusted using the feed rate magnification switch on the panel of the machine tool, but during thread machining, the feed rate magnification is prohibited.

4) Get ready function (G function)

(1) Quick point positioning command G00. The G00 command is used to control the tool to move rapidly from the current point to the target point in the point control mode. It is just quick positioning command and has no trajectory requirements. The rapid traverse speed is determined by the internal parameters of the system. The feed speed command F doesn't work together with the G00 command, which can be adjusted by the rapid feed rate magnification switch on the operation panel of the machine tool.

The programming format is

G00 X_ Y_ Z_,

where X, Y, and Z stand for the coordinate value of the target point for tool movement, and the value can be specified as absolute coordinate value or incremental coordinate value by G90 or G91.

G90 indicates that the block is programmed in absolute coordinates, and the coordinate values of the tool (or machine tool) movement position are given relative to the programming origin.

G91 indicates that the block is programmed in incremental coordinates, and the coordinate value of the tool movement position is relative to its previous position. The relative coordinates are related to the direction of movement.

As shown in Figure 2-2-1, the tool moves from point *A* to point *B* and the movement is programmed with G90 and G91 respectively. The blocks are as follows,

G90   G00   X130.0   Y120.0   F300;
G91   G00   X70.0   Y70.0   F300.

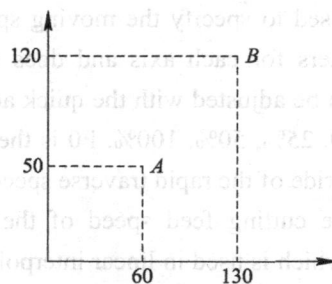

Figure 2-2-1   Tool Trajectory

When executing the G00 command, because each axis moves at its own speed, it cannot be guaranteed that each axis reaches the end point at the same time. The combined trajectory of the linked linear axis may not necessarily be a straight line, which varies according to the CNC system of the machine tool. Therefore, programmers should know the tool movement trajectory of the CNC system, and pay attention to whether the tool interferes with the workpiece and the fixture. On occasions that are not suitable for linkage, each axis can move in one way to avoid possible collisions during processing.

As shown in Figure 2-2-2 (a), the tool moves quickly from point A to point B, and is programmed with G90 and G91 respectively.

The linkage block of coordinate axis is as follows,

    G90   G00   X60.0   Y50.0; (Absolute coordinate programming)

    G91   G00   X – 40.0   Y – 70.0. (Incremental coordinate programming)

The single-action block of coordinate axis is as follows,

    G90   G00   X60.0; (Absolute coordinate programming)

                  Y50.0;

    G91   G00   X – 40.0; (Incremental coordinate programming)

                  Y – 70.0.

The single-action trajectory is shown in Figure 2-2-2 (b).

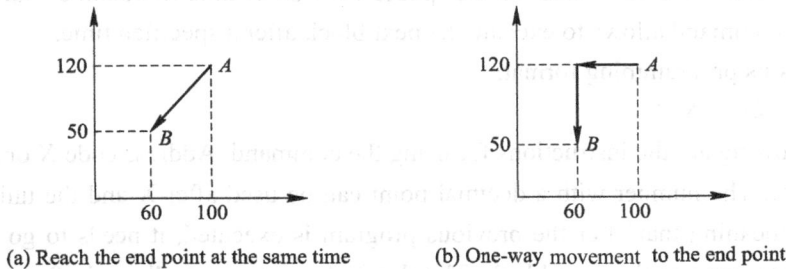

(a) Reach the end point at the same time   (b) One-way movement to the end point

Figure 2-2-2   Quick Point Positioning Tool Path

(2) Linear interpolation command G01. G01 is used to cut the curve contour in any slope approaching by straight-line segment. The tool is calculated by linear interpolation, and the movement speed is set by the feed function command F. Therefore, the F command must be present in the program when the command is executed.

Below is the programming format,

    G01 X_ Y_ Z_ F_.

Instructions for using the command are as follows.

① The coordinate value following G01 command can be formulated by G90 or G91 in absolute coordinate mode or incremental coordinate mode respectively.

② F command refers to modality command, it can be canceled with G00 command. If there is no F command in the block before G01, and if there is no F command in the current G01 block, the machine tool does not operate.

③ In the case that the additional linkage control function is selected, address of the fourth axis (A, B or C) can be used to replace X, Y or Z.

As shown in Figure 2-2-3, the tool starts linear interpolation from point A to point B, the feed speed is 200/min, and the G01 command is used for programming. The program is as follows,

　　　　G90　G01　X130.0　Y120.0　F200, or

　　　　G91　G01　X70.0　Y70.0　F200.

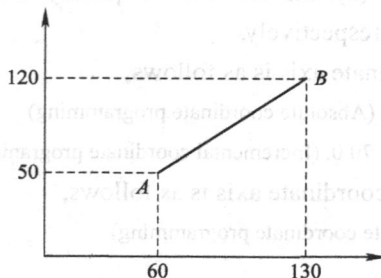

Figure 2-2-3　Interpolation Trajectory of Tools

5) Pause command G04

The G04 command can make the tool pause for a short time to obtain a round and smooth surface. This command allows to execute the next block after a specified time.

Below is its programming format,

　　　　G04　X (P).

The following are the instructions for using the command. Address code X or P refers to the time for pause. The number with a decimal point can be used after X and the unit is s, such as "G04 X5.0" meaning that after the previous program is executed, it needs to go through a 5 s pause, and then execute the next block. The decimal point is not allowed after address code P and the unit is ms, such as "G04 P5000" meaning that the time for a pause is 5 s.

6) Plane selection commands G17, G18 and G19

The plane selection command is used to select the plane for arc interpolation and tool compensation.

G17: Select the plane X-Y. In the program, X and Y are used to perform compensation calculation, and the coordinate value of Z axis is not affected by the compensation.

G18: Select the plane Z-X. In the program, Z and X are used to perform compensation calculation, and the coordinate value of Y axis is not affected by the compensation.

G19: Select the plane Y-Z. In the program, Y and Z are used to perform compensation calculation, and the coordinate value of X axis is not affected by the compensation.

7) Arc interpolation command G02 and G03

The arc interpolation command is used to control the tool to perform arc cutting at a given feed speed (specified by F) in the specified plane to process a contour of arc.

(1) Programming formats. The programming formats for G02 and G03 are of two modes.

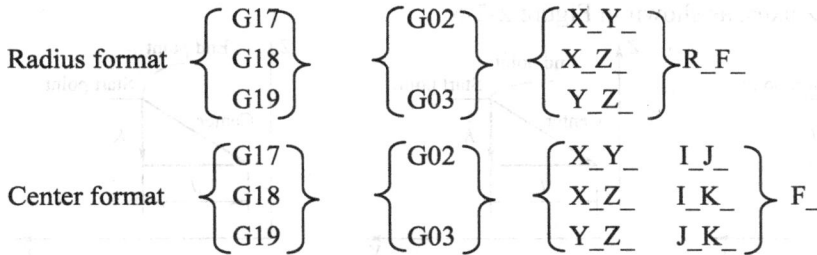

Radius format $\left\{\begin{array}{l} G17 \\ G18 \\ G19 \end{array}\right\}$ $\left\{\begin{array}{l} G02 \\ \\ G03 \end{array}\right\}$ $\left\{\begin{array}{l} X\_Y\_ \\ X\_Z\_ \\ Y\_Z\_ \end{array}\right\}$ R_F_

Center format $\left\{\begin{array}{l} G17 \\ G18 \\ G19 \end{array}\right\}$ $\left\{\begin{array}{l} G02 \\ \\ G03 \end{array}\right\}$ $\left\{\begin{array}{ll} X\_Y\_ & I\_J\_ \\ X\_Z\_ & I\_K\_ \\ Y\_Z\_ & J\_K\_ \end{array}\right\}$ F_

(2) Instructions for using the command.

① The commands related to arc machining are shown in Table 2-2-4.

Table 2-2-4   Instructions for Arc Interpolation Commands

Command	Specified contents		Meaning
G17	Specified plane		X-Y plane arc
G18			X-Z plane arc
G19			Y-Z plane arc
G02	Direction of rotation		Clockwise arc interpolation
G03			Counterclockwise arc interpolation
Two of the X, Y and Z (such as X_Y_)	End position G90 mode		Coordinates of the end position in the coordinate system of workpiece
	End position G31 mode		Coordinates of the end point relative to the start point
Two of the I, J and K (such as I_J_)	Distance from the starting point to the center of the circle		Coordinates of the center of the circle relative to the start point of the arc
R	Radius of the arc		Radius of the arc
F	Feed speed		Tangent speed of the arc

② Make a judgment of the clockwise and counterclockwise direction of the arc. The positive direction of the coordinate axis (such as the Z axis) perpendicular to the plane where the arc is located (such as the X-Y plane) is G02, and the counterclockwise direction is G03, as shown in Figure 2-2-4.

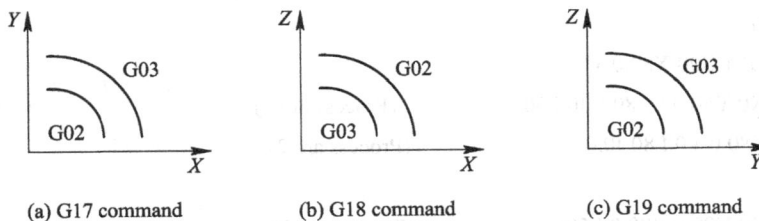

(a) G17 command   (b) G18 command   (c) G19 command

Figure 2-2-4   Judgment of Clockwise and Counterclockwise Direction of Arc

③ I, J, and K are the coordinate increments of the center relative to the start point of the arc on the X, Y, and Z axes, as shown in Figure 2-2-5.

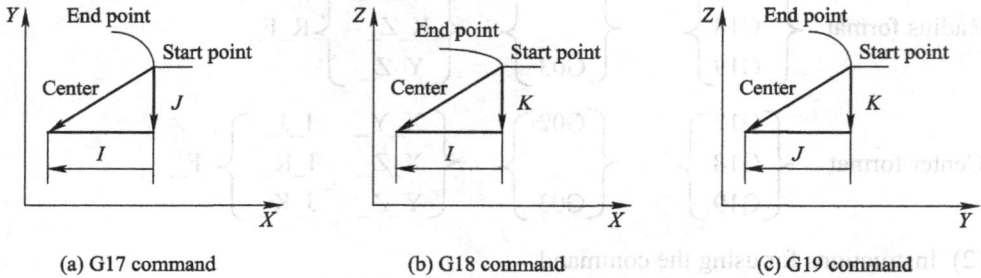

(a) G17 command　　　　　　(b) G18 command　　　　　　(c) G19 command

Figure 2-2-5　Determination of I, J and K axes value

④ For radius R of the arc, it is specified as follows. If the central angle $\alpha < 180°$, it is represented by "$+ R$", as the arc 1 in Figure 2-2-6; if $\alpha > 180°$, it is represented by "$-R$", as the arc 2 shown in Figure 2-2-6.

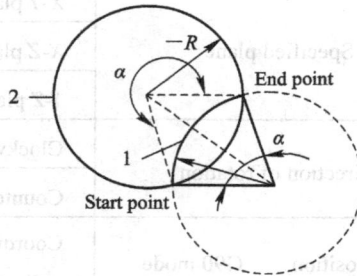

Figure 2-2-6　Positive and Negative Radius of Arc

⑤ Circle programming can use only the vector format rather than the radius format.

(3) Application example. As shown in Figure 2-2-7, the tool cuts at a tangential speed of 50 mm/min to complete the programming of the arc machining path (the tool is located above the point P first, and only performs trajectory movement). The program is as follows.

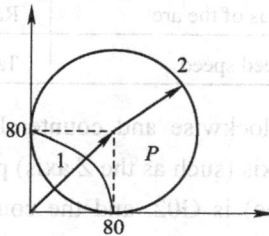

Figure 2-2-7　Application of Arc Interpolation

O0001;

G90 G54 G00 X80.0 Y0;

G03 X0 Y80.0 I −80.0 J0 F50;　　　　　(Process arc 1)

G02 X80.0 Y0 I 80 J0.0.　　　　　　(Process arc 2)

Or

G03 X0 Y80.0 R80.0 F50;　　　　　(Process arc 1)

G02 X80.0 Y0 R −80.0.　　　　　　(Process arc 2)

8) Polar coordination command G16/G15

G16 command enables the coordinate value to be input in the form of polar coordination radius and angle. G15 cancels the polar coordination command and causes the system to return to the rectangular coordination input.

(1) Programming format.

$$\left[\begin{matrix} G17 \\ G18 \\ G19 \end{matrix}\right] \left[\begin{matrix} G90 \\ \\ G91 \end{matrix}\right] G16 \qquad \text{(Start polar coordination command)}$$

G00 X_ Y_;   (Quick positioning of the tool to the position specified
                                         by polar coordination)

...

G15.   (Cancel polar coordination command)

In the formula, X and Y represent the polar diameter and polar angle in the polar coordination system.

(2) Instructions for using the command.

① In using polar radius, the plane should be determined firstly. In G18 plane, use Z axis to specify the polar diameter and X axis to specify the polar angle; in G19 plane, use Y axis to specify the polar diameter and Z axis to specify the polar angle. However, the Z axis generally isn't used in polar coordination, which is usually used in G17 plane.

② Polar coordination cannot be set in coordinate rotation and zoom mode, otherwise, the system will alarm.

③ The third axis that is not in the specified plane is independent of polar coordination.

④ The positive direction of the angle is the counterclockwise turning of the first axis of the selected plane, and the negative direction is the turning in the clockwise direction.

⑤ Both radius and angle can be specified with absolute coordinate command G90 or incremental coordinate command G91. The origin of the workpiece coordinate system is specified as the origin of the polar coordinate system, and the radius is measured from this point, as shown in Figure 2-2-8.

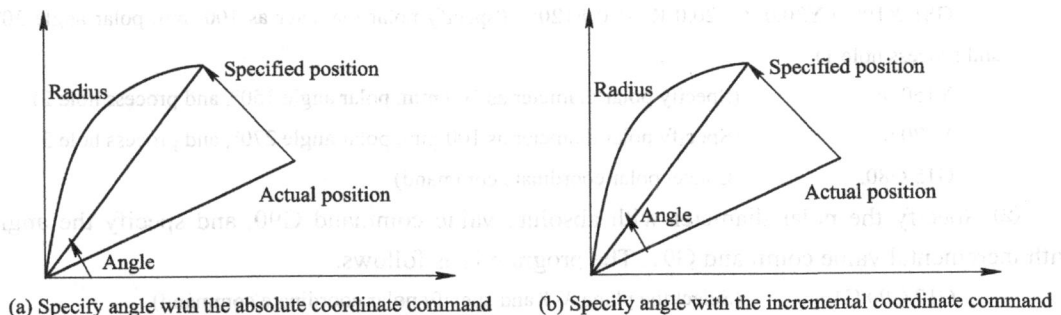

(a) Specify angle with the absolute coordinate command   (b) Specify angle with the incremental coordinate command

Figure 2-2-8   Use of G16 Command in G90 Mode

The current position is used as the origin of the polar coordination system, and the radius is measured from this point, as shown in Figure 2-2-9.

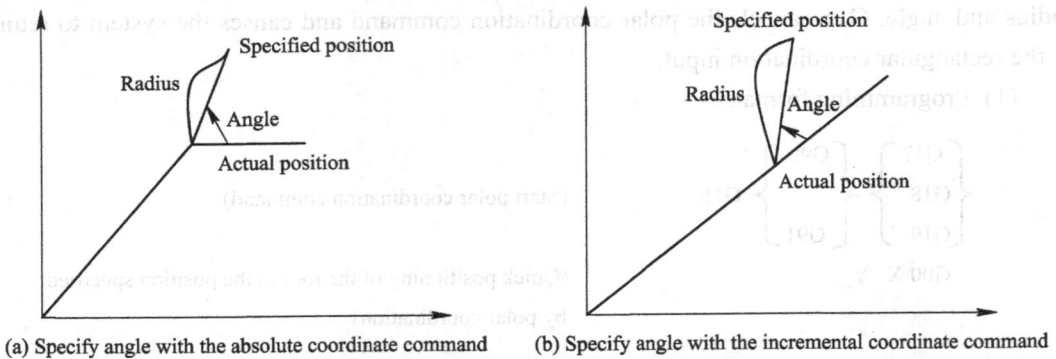

(a) Specify angle with the absolute coordinate command　　　(b) Specify angle with the incremental coordinate command

Figure 2-2-9　Use of G16 Command in G91 Mode

(3) Example for application. As shown in Figure 2-2-10, use G15/G16 instructions to program and process three circular holes 1, 2 and 3 in a circle with different radius.

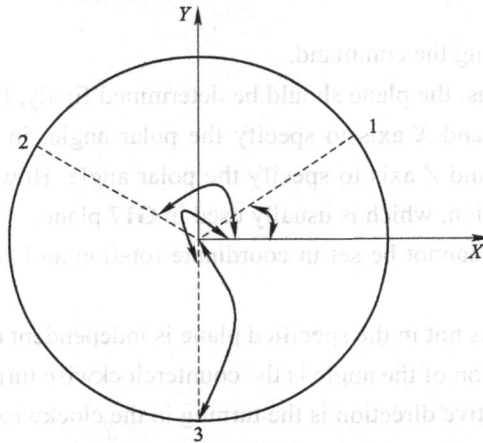

Figure 2-2-10　Example of Polar Programming

① To specify the angle and radius with absolute value command G90, the program is as follows.

　　　　G17　G90　G16;　　(Select the plane X-Y and specify polar coordinate command)

　　　　G81 X100.0 Y30.0 Z −20.0 R −5.0 F120;　(Specify polar diameter as 100 mm, polar angle 30°, and process hole 1)

　　　　Y150.0;　　　　　　(Specify polar diameter as 100 mm, polar angle 150°, and process hole 2)

　　　　Y270.0;　　　　　　(Specify polar diameter as 100 mm, polar angle 270°, and process hole 3)

　　　　G15 G80.　　　　　(Cancel polar coordinate command)

② Specify the polar diameter with absolute value command G90, and specify the angle with incremental value command G91. The program is as follows.

　　　　G17 G90 G16;　　　(Select the plane X-Y and specify polar coordinate command)

　　　　G81　X100.0　Y30.0　Z −20.0　R −5.0　F120.0;　(Specify polar diameter as 100 mm, polar
　　　　　　　　　　　　　　　　　　　　　　　　　　　angle 30°, and process hole 1)

G91   Y120.0;   (Specify polar diameter as 100 mm, incremental angle 120°, and process hole 2)

Y 120.0;   (Specified pole diameter 100 mm, incremental angle 120°, and process hole 3)

G15 G80.   (Cancel polar coordinate command)

It should be noted that G81/G80 in the program is the start command/cancel command for a drilling cycle, and R is the coordinate of the starting point of tool (the upper surface of the workpiece is the coordinate origin).

9) Tool compensation command

The CNC programming of the parts assumes the tool center (or tool tip) moves relative to the workpiece. In milling the contour of the workpiece with a milling cutter, because the tool always is in a certain radius, the movement path of the tool center does not coincide with the actual contour of the part to be processed. As shown in Figure 2-2-11, the thick solid lines are the contour of the part to be processed, and the double-dotted line is the path of the tool center. If the trajectory of the tool center is directly programmed, it needs to be calculated by the size of the workpiece contour and the tool radius. The calculation is quite complicated and must be redone and the program must be modified if the size of tool changes, which is very inconvenient.

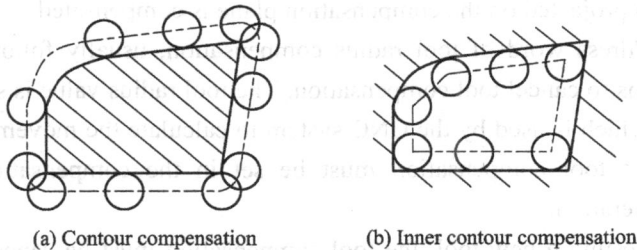

(a) Contour compensation   (b) Inner contour compensation

Figure 2-2-11   Tool Radius Compensation

In order to simplify the CNC programming of the parts and make it as independent as possible of the shape and size of the tool, the CNC system generally has the tool compensation function. There are usually three types of tool compensation, tool radius compensation, tool length compensation and tool wear compensation.

For Tool radius compensation, commands are introduced as follows.

(1) Tool radius compensation command G41/G42 and G40. When the CNC system performs the tool radius compensation function, the tool radius is stored in the compensation memory of the CNC system. CNC programming only needs to be carried out according to the contour of the workpiece. The CNC system automatically calculates the tool center trajectory for the tool to deviate from the contour of the workpiece by the value of a radius, that is, the tool radius compensation is processed. The program is as follows.

G41/G42   G01/G00   X_Y_. (Perform left/right tool compensation)

G40 G01/G00   X_Y_. (Cancel tool compensation)

(2) Instructions for using the command.

① G17, G18 and G19 are selection commands for plane compensation. Before performing tool radius compensation, the working plane must be specified with G17, G18 or G19 command.

The command for plane selection must be switched when in the mode of compensation cancellation; otherwise, an alarm is sent.

② G41 is the left compensation of tool radius. Looking along the direction in which the tool moves forward, the tool is located on the left side of the part to perform compensation (left compensation), as shown in Figure 2-2-12(a); G42 is the right compensation of tool radius, and the tool is located on the right side of the part for compensation, it can be regarded as along the tool's forward direction (right compensation), as shown in Figure 2-2-12(b); G40 cancels the tool radius compensation and is used to cancel the G41/G42 commands. G40/G41 and G42 are modal codes, which can be cancelled by each other.

③ G40/G41 and G42 can only be combined with G01 or G00 to complete the establishment and cancellation of tool radius compensation, rather than with the arc interpolation commands G02 or G03.

④ X, Y (or X, Z and Y, Z) is the coordinate command of the target point of G01 or G00 motion, that is, the end point of the establishment or cancellation of tool compensation. The coordinate axis must correspond to the axis in the specified plane. Under multi-axis linkage control, the tool path projected on the compensation plane is compensated.

⑤ D is the address word of tool radius compensation, usually followed by two digits, D00–D99. D00 means to cancel tool compensation. The tool radius value is stored in the D code as the offset value, which is used by the CNC system to calculate the movement path of the tool center. The value of tool compensation must be set in the compensation memory before machining or trial operation.

⑥ Before requesting a new tool, the tool compensation must be cancelled, otherwise, an alarm is sent.

(3) The process of tool radius compensation. The process of tool radius compensation is divided into three stages, establishment, operation and cancellation, as shown in Figure 2-2-12. Before establishing tool radius compensation, the tool should be at a proper distance from the part contour (generally larger than the value of tool radius compensation), and should be coordinated with the selected cut-in point and feed method to ensure the effectiveness of tool radius compensation. The end point of the tool for tool radius compensation (cancellation) should be placed after the tool cuts out the workpiece, otherwise, a collision will occur.

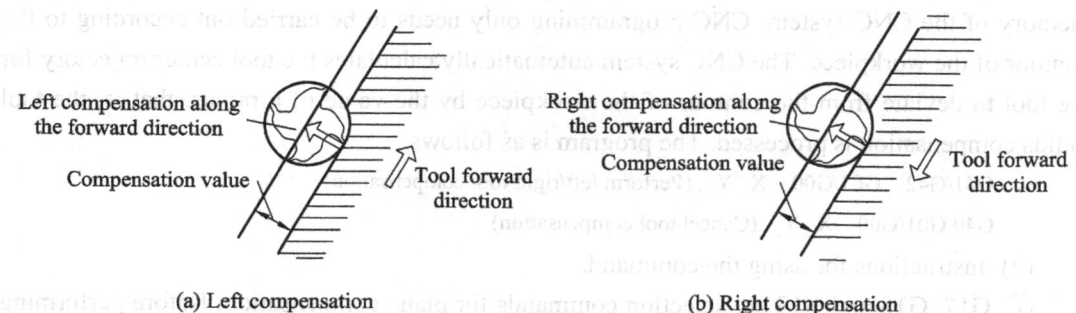

Left compensation along the forward direction

Compensation value

Tool forward direction

Right compensation along the forward direction

Compensation value

Tool forward direction

(a) Left compensation　　　　(b) Right compensation

Figure 2-2-12　Process of Tool Radius Compensation

The establishing block for tool radius compensation should be completed before cutting into the workpiece; the canceling block for tool radius compensation should be completed before cutting out of the workpiece, otherwise, overcutting will arise.

The tool radius compensation section for linear length establishment and linear length canceling should be greater than the compensation value, otherwise, the system alarms during machining.

(4) Example for application. In the plane *X-Y*, use the radius compensation command to complete the programming of contour machining, as shown in Figure 2-2-13.

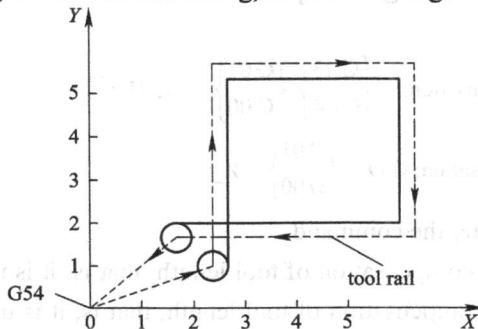

Figure 2-2-13   Example of Tool Radius Compensation

The programming is as follows.

    O0002;

    ...

    N020 G90 G54 G00 X0 Y0 M03 S500 F50;

    N030 G00 Z50.0;   (Starting height)

    N040 Z10.0;   (Safe height)

    N050 G41 X20.0 Y10.0 D01;   (Tool radius compensation)

    N060 G01 Z-10.0;   (Put down cutter, with a cutting depth of 10mm)

    N070 Y50.0;

    N080 X50.0;

    N090 Y20.0;

    N100 X10.0;

    N110 G00 Z50.0;   (Pull up cutter to the starting height)

    N120 G40 G01 X0 Y0 M05;   (Cancel tool radius compensation)

    N130 M30.

(5) Examples of tool radius compensation.

① In addition to making programming more convenient, tool radius compensation can also be applied in other ways. In actual processing, the function of tool radius compensation can be used to perform rough or finish machining with the same program and the same size of tool. Namely, in rough machining, the value of tool radius compensation = tool radius + finish allowance; in finish machining, the value of tool radius compensation = tool radius + correction amount.

② If tool diameter changes due to wear, regrind and tool replacement, it is not necessary to modify the program and just input the changed tool radius in the parameter setting.

For Tool length compensation, commands are introduced as follows.

(1) Tool length compensation command G43/G44 and G49. In setting coordinate system of the workpiece, make the reference plane of the taper hole of spindle coincide with the theoretical surface on the workpiece. In using each tool, the machine tool can be raised at a distance according to the length of the tool, so that the tool tip is right on the surface of the workpiece. The height is the value of tool length compensation. G codes for tool length compensation are G43/G44 and G49.

$$\text{Perform tool compensation:} \quad \begin{Bmatrix} G43 \\ G44 \end{Bmatrix} \begin{Bmatrix} G01 \\ G00 \end{Bmatrix} \quad Z\_H\_;$$

$$\text{Cancel tool compensation: G49} \quad \begin{Bmatrix} G01 \\ G00 \end{Bmatrix} \quad Z\_;$$

(2) Instructions for using the command.

① G43 is the positive compensation of tool length, that is, it is used to realize the positive offset; G44 is the negative compensation of tool length, that is, it is used to realize the negative offset. As shown in Figure 2-2-14, the drilling bit uses G43 command to compensate the offset of tool 1 in positive direction, and the milling tool uses G43 command to compensate the offset of tool 2 in the upward direction. G49 cancels the G43/G44 commands for tool compensation.

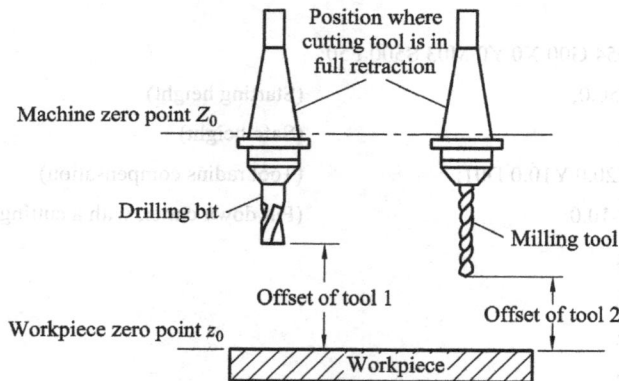

Figure 2-2-14　Example of Tool Length Compensation

② In G code, Z follows with the end point value as the compensation value, and H points to the memory address of the compensation value of tool length.

③ In using G43/G44 commands, whether for absolute coordinate programming or incremental coordinate programming, the coordinate value of Z-axis movement specified in the program must be calculated with the offset in the register specified by H, add the offset when use G43 and subtract the offset when use G44.

④ G43/G44 and G49 are modal codes and can be canceled by each other.

For Tool wear compensation, it is processed as follows:

In the cutting process, on the one hand, the tool cuts off the chips, on the other hand, the tool itself gradually wears out. If the tool wears out during the machining process, the machining accuracy is reduced. As the cutting time passes, the amount of tool wear continues to increase, which inevitably leads to changes in the size of the workpiece, thereby affecting the processing quality of the workpiece. In order to ensure the machining accuracy of the parts, tool wear compensation should be carried out.

Commonly used compensation methods include manual compensation method and automatic compensation method. Automatic compensation method is further divided into online measurement automatic compensation method and macro program automatic compensation method.

3. Manual Programming of Plane Milling Program

With the CNC machining center of the FANUC system and under the conditions of the commonly-used fixture, please complete the programming of the upper plane milling of the aluminum material as shown in Figure 2-2-15.

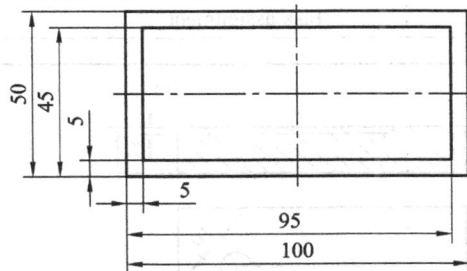

Figure 2-2-15　Upper Plane Milling

1) Determination of the processing path of parts

The processing path (see Figure 2-2-16) of the parts refers to the path and direction of the movement of the tool position relative to the processed part in the processing of the CNC machine tool.

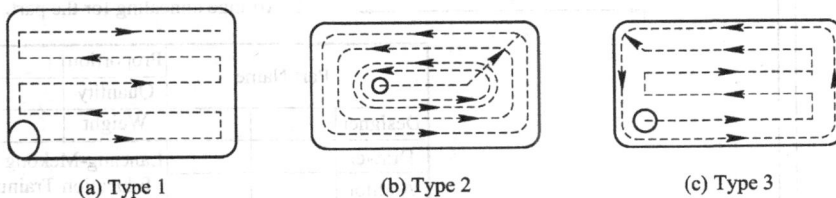

(a) Type 1　　　　(b) Type 2　　　　(c) Type 3

Figure 2-2-16　Processing Path

Below are the principles for determining the processing path.

① The processing accuracy and surface roughness of parts should be ensured;

② The processing path should be shortened as much as possible to reduce the time of empty travel of tools;

③ The numerical calculation should be simple and the number of blocks should be small to reduce the programming workload.

2) Starting point (aka tool setting point of the program)

The starting point of the tool is the starting point of the tool relative to parts movement when in machining parts on CNC machine tools.

Below are the principles for choosing the tool starting point.

① It should be easy to handle mathematical processing and simplify programming;

② It should be easy to align on the machine tool, and be convenient to check during processing;

③ The processing error should be small.

4. Skills Training

Complete the task sheet as Table 2-2-5 based on what you have learned above.

Table 2-2-5 Task sheet

Task name	Manual programming of simple parts		
Class		Group No.	
Task assignment			
Drawings of parts	Technical specifications : 1. Sharp-corner bevel edge, seamed-corner bevel edge C2. 2. Arrange annealing for the part.		
Programming			

Part Name	Proportion		01
	Quantity		
Designer	Weight		45#
Drawer	Lancang-Mekong Vocational Education Training Center		
Auditor			

Module 3

Basic Operation of Machining Center

Task 1　Use of the Operation Panel of Machining Center

【Knowledge objectives】

　　(1) Know the operation panel of the machine tool of machining center (VMC850).

　　(2) Know the control panel of FANUCOi-f CNC system.

【Ability objectives】

　　(1) Master the use of the operation panel of machine tool.

　　(2) Master the use of the operation panel of FANUCOi-f CNC system.

1.　The Operation Panel of the Machine Tool of FANUCOi-f CNC System

　　The display panel and operation panel of FANUC and other CNC systems are shown in Figure 3-1-1 and Figure 3-1-2 respectively.

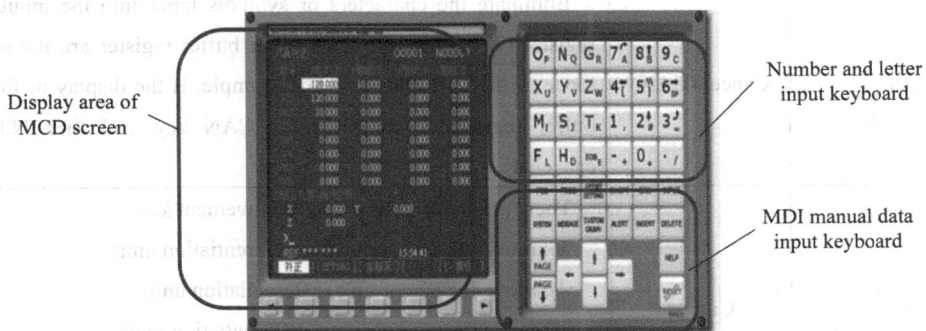

Display area of
MCD screen

Number and letter
input keyboard

MDI manual data
input keyboard

Figure 3-1-1　Display Panel of Machine Tool

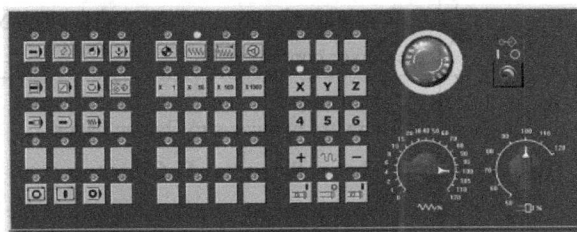

Figure 3-1-2　Operation Panel of Machine Tool

2. The Control Panel of FANUCOi-f CNC System

1) Description of the control panel of CNC system

The whole panel is divided into three parts, the upper left corner of Figure 3-1-1 being the screen display area, the right being the editing area of MDI manual program and the display panel of machine tool. In this area, you can edit the program, display the position, change the tool and set the parameters, etc. Figure 3-1-2 is the operation panel of the machine tool. This interface is used to set the auxiliary functions of the machine tool, such as forward and backward rotation of spindle, coolant switch, feed rate adjustment, etc.

2) Function Description of keys

The function description of keys is shown in Table 3-1-1.

Table 3-1-1　Function description of Keys on the Display Panel of Machine Tool

Keys	Name	Function
RESET	Reset	Cancel the alarm and reset the CNC
Letter/number keys	Letter / number keys	Enter numbers, letters and other characters
INPUT	Input (IN)	Input data such as parameters and compensation amount
CAN	Cancel (CAN)	Eliminate the characters or symbols input into the input buffer register (The contents of the input buffer register are the same as those of the CRT display). For example, If the display of the input buffer register is N0001, press CAN key, and then N0001 is canceled
Cursor movement key	Cursor movement key	There are four kinds of cursor movement keys: ↓ : move the cursor down a differentiation unit; ↑ : move the cursor up a differentiation unit; → : move the cursor right a differentiation unit; ← : move the cursor left a differentiation unit; Continuously press the up and down keys of the cursor to move the cursor continuously
PAGE / PAGE	Page	There are two ways to change pages: ↓: change the pages of LCD screen in the forward direction; ↑: change the pages of LCD screen in the backward direction

Table 3-1-2 shows the function description of keys on the operation panel of the machine tool.

Table 3-1-2　Function Description of Keys on the Operation Panel of Machine Tool

Keys	Name	Function
	Emergency stop	Stop in case of emergency
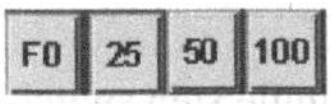	Fast shift override	It is invalid to adjust rapid override, feed override, manual speed, spindle override if override is prohibited
	Program start	In automatic or MDI mode, the program starts to run
	Program pause	In running the program, press this button to pause its running, and then press cycle start key to start where the program is paused
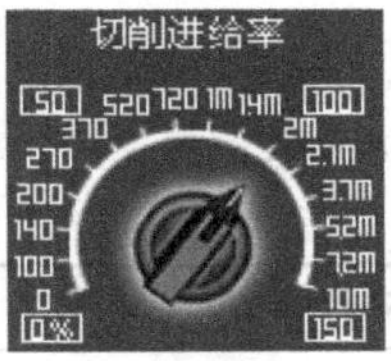	Cutting feed rate	After move the cursor to this knob, click the left or right mouse button to adjust the cutting feed rate
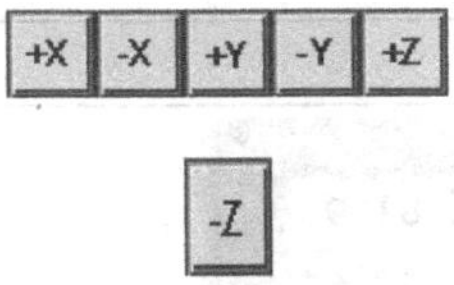	+X/ −X/ + Y/ − Y/ + Z/ − Z	Move in +X/ − X/ + Y/ − Y/+Z/ − Z direction
	Spindle deceleration, spindle 100%, spindle acceleration	Control the percentage of spindle speed by rotating the designated gear

Notes:

In manual/continuous processing or when the tool setting requires precise adjustment of the machine tool, the machine tool can be process in handwheel mode.

① Click the button in the operation panel to switch to the handwheel mode;

② Select the axis to be moved by knob 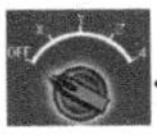 . By adjusting the override knob  on the operation panel, click the left or right mouse button on the handwheel  to precisely control the machine tool, where X1 represents 0.001 mm, X10 represents 0.01 mm and X100 represents 0.1 mm.

③ Click the button  to control the spindle to move forward, stop and reverse respectively.

The button description is shown as in Table 3-1-3.

Table 3-1-3   Button Description

Keys	Name	Keys	Name
	Auto run		Reference point return
	Edit mode		Manual mode
	MDI input method		Single step switch
	Single block		One-hand / handwheel movement amount
	Empty run		Manual axis selection

3. Skills Training

Complete the task sheet as Table 3-1-4 based on what you have learned above.

Table 3-1-4   Task sheet

Task name	The operation panel of machining center (VMC850)		
Class		Group No.	
Task assignment			
The operation panel			
Precautions for operation			

Task 2   Basic Operation Requirement of Machining Center

【Knowledge objectives】

(1) Understand the method for the start and shutdown of machining center and its initialization (return to reference point).

(2) Master the input and edition functions of machining center.

(3) Learn how to install fixtures (flat vise, vise, etc.) in machining center.

【Ability objectives】

(1) Master the method for the start and shutdown of machining center and the initialization of machining center (return to the reference point).

(2) Master the input and editing method of machining center.

(3) Master the basic operation method of machining center.

(4) Master the installation method of fixtures.

1. Startup/Shutdown and Initialization of Machining Center (Return to Reference Point)

1) Startup of machine

(1) Precautions before startup.

① The operator must be familiar with the performance and the operation method of the CNC machine tool. The machine can be operated only with the consent of the machine tool management personnel.

② Before powering on the machine, check whether the voltage, air pressure and oil pressure meet the working requirements.

③ Check whether the movable parts of the machine tool are in a normal state.

④ Check whether the workbench is offside and over limit.

⑤ Check whether the electrical components are firm and whether there are wires falling off.

⑥ Check whether the ground wire of the machine tool is reliably connected with that of the workshop.

⑦ Do not turn on the main power switch until the preparatory work has been completed.

(2) Precautions for the starting process.

① Operate in strict accordance with the starting sequence in the machine manual.

② Generally, after the startup, you must first return to the machine tool reference point to establish the machine tool coordinate system.

③ Allow empty running of the machine for about 15 minutes after its startup, so that the machine can reach a balanced state.

④ After the shutdown, you must wait for more than 5 minutes before restarting the machine. Do not start up and shut down the machine frequently except on special circumstances.

(3) Precautions for debugging.

① If it is the first trial cutting, operate the machine in empty running.

② Install and adjust the fixture according to process requirements, and remove iron chips and debris from each positioning surface.

③ Install the workpiece according to the positioning requirements and ensure its positioning to be correct and reliable. The workpiece should not be loose during processing.

④ Install the tool to be used. The tool position number on the tool bank must be strictly consistent with the tool number in the program.

⑤ Set the tool according to the programming origin on the workpiece and establish a workpiece coordinate system. If multiple tools are used, once decide on the position of first tool, the length compensation and the tip position compensation of other tools should be carried out.

⑥ Set the tool radius compensation.

⑦ Confirm that the coolant output is smooth and its flow is sufficient.

⑧ Check again whether the established workpiece coordinate system is correct.

⑨ Workpieces cannot be processed until the above steps are implemented.

2) Shutdown of machine

(1) Precautions before shutdown.

① The operator must be familiar with the performance of and the method for the operation of the CNC machine tool. The machine tool can be operated only with the consent of the machine tool management personnel.

② Check whether the workbench is offside and over limit.

③ After make above preparations, shutdown the main power switch.

(2) Precautions for the shutdown process.

① Operate in strict accordance with the shutdown sequence in the machine manual.

② Under normal circumstances, before the shutdown, you must first remove the milling machine table to the middle of the guide rail, and remove the carriage of the lathe to the side of the tailstock to prevent the guide rail from deforming.

③ After shutdown, you must wait for more than 5 minutes before restarting. Do not start and shut down the machine frequently except on special circumstances.

(3) Precautions after shutdown.

① Cut off the main power supply.

② Clean up the work site.

3) Reset

When the system is powered on and enters the software operation interface, the working mode of the system is "Emergency Stop". To control the operation of the system, turn left and pull up the "Emergency Stop" button in the upper right corner of the console to reset the system and turn on the servo power. By default, the system enters the "Return to Reference Point" mode, and the working mode of the software operation interface becomes "Return to Zero".

4) Return to the reference point

After the CNC machine tool is powered off, the system automatically loses the position memory of the coordinate axis, so the machine tool has to start to return to zero first. Return each coordinate axis of the machine tool to a fixed position (zero point of the coordinate system of the machine tool). "Return to Reference Point" is one of the most important functions in the operation of CNC machine tools, which directly affects the tool compensation, gap compensation, axial compensation, accuracy compensation and processing quality of CNC machine tools.

The reasons for the failure in returning to the reference point are as follows.

① If the position of the reference point return switch is improper, the real zero point pulse appears during the zero return deceleration process, the system queries speed is less than the movement speed of the coordinate axis, the current zero point pulse signal is lost, and the next zero point pulse appears before the deceleration and stops at the place where the stopped position exceeds zero point;

② Drift phenomenon occurs in the movement gap of the mechanical structure, and the stopped position deviates from zero point for a small distance;

③ The stopped position deviates from zero point due to incorrect parameter settings (such as displacement counter, zero return operation speed, grid shielding amount and zero point offset, etc.). What needs to be checked are the zero point switch, the position of the zero-axis pressing block, the state of the mechanical structure gap and the zero-return parameters.

The premise of controlling the movement of the machine tool is to establish its coordinate system. Therefore, after the system is powered on and reset, the operation of returning each axis of the machine tool to the reference point should be carried out first. The method is as follows.

① If the current working mode displayed by the system is not the zero-return mode, press the "Zero Return" button on the control panel to ensure that the system is in the "Zero Return" mode.

② Press the "+ X" ("reference point direction" is "+") or "−X" ("reference point direction" is "−") button according to the X axis machine parameter "reference point direction". After the X axis returns to the reference point, the indicator light in the "+ X" or "−X" button lights up.

③ In the same way, use the "+ Y" or "−Y" button to return the Y axis to the reference point.

④ In the same way, use the "+ Z" or "−Z" button to return the Z axis to the reference point.

5) Precautions

① After each power-on, the operation of returning to the reference point of each axis must be completed first, and then enter other operation modes to ensure the accuracy of the coordinates of each axis.

② Press the axial selection buttons X, Y, and Z at the same time to return the X, Y, and Z axes to the reference point at the same time.

③ Before returning to the reference point, make sure that the zero-return axis is located to the opposite side of the "reference point direction" of the reference point (if the X reference point return direction is negative, ensure that the current position of the X axis is at the positive side of the reference point before returning to the reference point), otherwise, the axis should be moved manually until this condition is met.

④ In the process of returning to the reference point, if an overtravel occurs, press and hold the "Overtravel Release" button on the control panel and manually move the axis in the opposite direction to quit the overtravel state.

2. Input and Editing of Machining Center

1) Create a new program

The display panel of machine tool is shown as in Figure 3-2-1.

① Select the working mode button as "EDIT";

② Press the MDI function key "PROG";

③ Input the address O, input the program number (such as 0001), and press the "INSERT" key;

④ Press the "EOB" key and press the "INSERT" key again to complete the insertion of the new program "O0001".

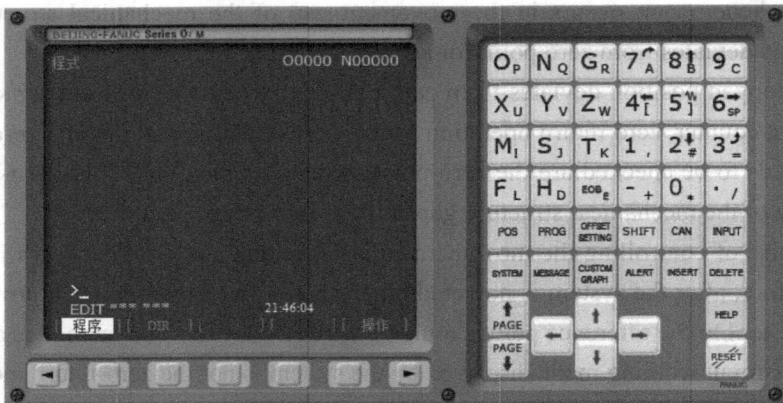

Figure 3-2-1   Display Panel of Machine Tool

2) Request for the program stored in the memory

① Select "EDIT" for the mode button;

② Press the MDI function key "PROG", input the address O, and input the program number to be requested, such as 0001;

③ Press the cursor movement key to complete the request of program "O0001".

3) Deletion of a program

① Select "EDIT" for the mode button;

② Press the MDI function key "PROG", input the address O, and input the program number to be deleted, such as "0001";

③ Press the "DELETE" key to complete the deletion of a single program "O0001".

If you want to delete all the programs in the memory, just press "DELETE" after inputting "0-9999" and then press the soft key "EXEC" (in the left of display panel) to delete all the programs in the memory.

If you want to delete a specified program, as long as you input "OXXXX", press the "DELETE" key and then press the soft key "EXEC", then the program is deleted.

4) Deletion of a block

① Select "EDIT" as the working mode button;

② Use the cursor movement keys to retrieve or scan the block address N to be deleted, and press the "EOB" key;

③ Press the "DELETE" key to delete the block where the cursor is currently located.

If you want to delete more than one block, use the cursor movement key to retrieve or scan the start address N (such as 10) of the block to be deleted, input the address N and the last block number (such as 60), press the "DELETE" key, and the blocks of N10−N60 are deleted.

5) Retrieval of a block

The retrieval function is mainly used in the automatic operation process. The retrieval process is as follows.

① Press the mode selection button "EDIT";

② First press the "MDI" function key to display the program screen, input the address O and the block number to be retrieved, and then press the "PROG" key to retrieve the block segment.

6) Operation of a program word

① Scan the program word. Select "EDIT" with the mode button, press the left "←" or right "→" move key of the cursor, and the cursor moves one address word to the left or right on the screen. Press the up " ↑ " or down " ↓ " move key of the cursor, the cursor moves to the beginning of the previous or next block. Press the "PAGE ↑ " key or "PAGE ↓ " key, the cursor pages backward or forward.

② Skip to the beginning of the program. In the "EDIT" mode, press the "RESET" key for the cursor to skip to the head of program.

③ Insert a program word. In the "EDIT" mode, scan the word before the position to be inserted, input the address and data to be inserted, and press the "INSERT" key.

④ Replacement of words. In the "EDIT" mode, scan the words to be replaced, input the new address words and data, and press the "ALTER" key.

⑤ Deletion of words. In the "EDIT" mode, scan the words to be deleted, and press the "DELETE" key.

⑥ Cancellation of words during input. During the input of program characters, if the current character input is incorrect, press the "CAN" key once to delete a currently input character.

3. Basic Operation Instruction of Machining Center

① After the machine tool is started up, exchange the tool at the No. 1 tool position in the tool bank on the spindle of the machine tool. After the tool change is completed, enter the spindle speed of the machine tool through the edit area on the display screen of the machine tool, and set the spindle speed to 500 r/min, to rotate forward. The editing instructions are as follows.

G91G28Z0;

M6T1;

M3S500.

② Complete the movement of the machine tool spindle and worktable in handwheel mode, and find the changes in the coordinates of the three coordinate systems. First, press the handwheel function button to select three magnification adjustment modes, ×100 (representing 0.1 mm for one rotation), ×10 (representing 0.01 mm for one rotation), and ×1 (representing 0.001 mm for one rotation). Rotate the handwheel to make the machine tool spindle move up and down, and make the worktable to move left and right. Under handwheel mode, the spindle stop function, forward and reverse rotation function (previously having a rotational speed), cutting fluid stop button and other functions of the machine tool are all valid.

③ Under manual mode, functions such as the rapid movement of the machine tool spindle (previously having a rotational speed) and the cutting fluid stop button can be realized. The rapid movement speed of the machine tool is determined by the rapid override adjustment.

④ If the machine tool returns to the origin, the operation sequence is to first return to zero in the Z direction, and then return to zero in the X/Y direction.

⑤ Having familiarized yourself with the above operations, pay attention to the protruding length of the tool and the storage address of the tool bank when clamping the workpiece and the tool (be cautious that when clamping in the editing mode, do not operate the machine tool).

4. Installation of In-machine Fixtures (Flat Vise, Vise, etc.) in Machining Center

Use the dial indicator to accurately calibrate the vise. In calibration, make the magnetic table base be attached to the rail surface of the beam or the spindle of the vertical milling head, and install the dial indicator so that the measuring rod of the dial indicator is perpendicular to the plane of the fixed jaw. The measuring contact touches the plane of the jaw, and the measuring rod is compressed by about 0.3−0.5 mm. Move the workbench longitudinally and observe the dial indicator readings. If the number is consistent within the whole length of the fixed jaw, then the fixed jaw is parallel to the feed direction of the worktable, so as to obtain a good position accuracy during processing as shown in Figure 3-2-2.

Figure 3-2-2　Calibrate Accuracy of Vise with Dial Indicator

5. Skills Training

Complete the task sheet Table 3-2-1 based on what you have learned above.

Table 3-2-1   Task sheet

Task name	Basic operation of machining center	
Class		Group No.	
Task assignment		
Precautions for startup and shutdown of the machining center		
Methods for calibrating the vises		

Task 3   Inputting and Editing of CNC Machining Program

【Knowledge objectives】

(1) Learn the function of the CNC system for program editing and the method for using the MDI keyboard.

(2) Learn the editing, inputting and graphic simulating of the program for part processing.

【Ability objectives】

(1) Master the editing mode and the method for using the MDI mode of machine tool.

(2) Master the methods for program inputting and graphic simulating.

1. Function of CNC System for Program Editing and Method for Using the MDI Keyboard

1) Steps for manual inputting

① Set the machine tool in MDI working mode;

② Press the "PROG" program key;

③ Press the soft key in MDI screen, the processing program name "O0000" is displayed automatically;

④ Input the test program, such as "M03S800";

⑤ Press the "Cycle Start" key to run the test program;

⑥ Stop the operation in case of M02 or M30 command, or press the "RESET" button to end the operation.

2) Instructions for manual input

① MDI manual input program cannot be stored.

② Having pressed the "Cycle Start" key, the running block cannot be edited. After the program have been executed, the content of the input area is still reserved. When the "Cycle Start" key is pressed again, the machine runs again.

CRT is a display screen, which is used to display related data. Users can see the feedback information of CNC system from the screen. The MDI panel is where users input the CNC commands, which is the main input method of the CNC system. The position of each key on the MDI panel is shown in Figure 3-3-1.

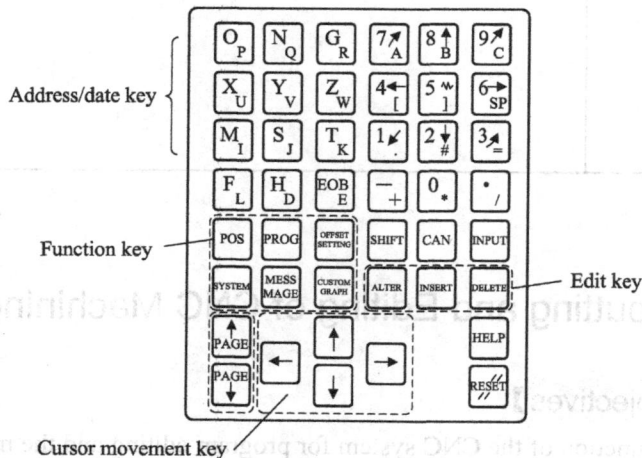

Figure 3-3-1   MDI Operation Panel

Table 3-3-1 gives a detailed description of each key on the MDI panel.

Table 3-3-1　Detailed Description of Keys on the MDI Panel

Serial number	Key name	Detailed description
1	RESET	Press this key, the CNC system will be reset or the alarm mode is cancelled
2	HELP	Provide help when you don't understand the operation of MDI panel
3	Soft key	Soft key provide different functions according to the change of screen interface, it is normally at the bottom of the screen
4	Address key and Number key O_P　$7^{\wedge}_{\wedge}$	Press the address and character key, you can input letters, numbers and other characters
5	SHIFT	SHIFT key helps to transfer functions of one key with two functions. When a letter on the lower right corner is input, a special character "^" will appear on the screen
6	INPUT	First press a character or number key, then press the INPUT key, the data is input into the buffer area and display on the screen. And press the INPUT key can help to copy the data to configuration register. INPUT key in MDI panel is in equivalence to INPUT key in soft key.
7	CAN	Press the CAN help to delete the last character or signal going into the buffer. For example: when the input data is :N001X100Z_, once press the CAN key, the character Z cancelled and the screen display is N001X100_
8	ALTER, INSERT, DELETE	Press the left keys for program editing
9	POS, PROG	Press the POS and PROG keys, alter the display screen with different functions, you can refer below paragraph for more detailed introduction about their functions

Serial number	Key name	Detailed description
10	Cursor move key	Move the cursor in four directions
11	[PAGE ↑] [PAGE ↓]	Turn over page up and down

2. Editing, Inputting and Graphic Simulating of Part Processing Program

If you input the CNC program manually, you should carefully check the program for grammar errors. However, if a logic error occurs in the program (the G code instruction is correct, but its statement does not meet requirements), it is impossible to detect it. Like the actual machine tool, the CNC machining simulation system also provides the function of tool path display. With this function, you can see the tool running trajectory of the program. The operation steps for displaying the tool running trajectory are as follows.

As shown in Figure 3-3-2, use the mouse to direct the mode selection knob to "Auto" (P1 in the Figure 3-3-2), press the "PROG" button (P2 in the Figure 3-3-2) in the system panel, then press the "CUSTOMGRAPH" button (P3 in the Figure 3-3-2), the display area of machine tool turning black, then press the cycle start button (P4 in the Figure 3-3-2) on the operation panel, and you can observe the running track of the CNC program. At this time you can also observe the three-dimensional operation track in all directions by the dynamic rotation, dynamic zoom, dynamic translation and other means in the "view" menu. The running trajectory is shown in Figure 3-3-2.

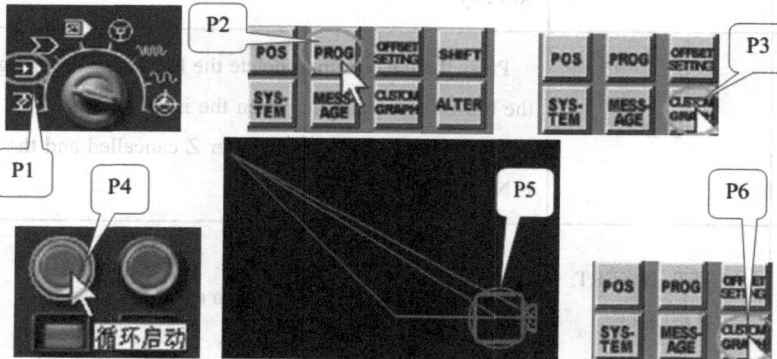

Figure 3-3-2   Check on Tool Running Trajectory

If the running trajectory of the tool is different from the assumed one, it means that the program is wrong. You can return to the editing mode of the program and correct the errors in it until there are no errors in the running trajectory.

3.　Skills Training

Complete the task sheet as Table 3-3-2 based on what you have learned above.

Table 3-3-2　Task sheet

Task name	Program input of a simple part of machining center		
Class		Group No.	
Task assignment			
Drawings for inputting	Technical specifications : 1. Sharp-corner bevel edge, seamed-corner bevel edge C2. 2. Arrange annealing for the part.		
		Proportion	01
	Part Name	Quantity	
	Designer	Weight	45#
	Drawer	Lancang-Mekong Vocational Education Training Center	
	Auditor		
Process of inputting			

Task 4　Tool Setting of Machining Center

【Knowledge objectives】

(1) Learn how to use the tool bank of machining center.

(2) Understand the use of the edge finder and the setting of the workpiece coordinate system.

(3) Understand the setting of tool shape compensation.

【Ability Objectives】

(1) Master the use of tool bank.

(2) Master the method for calibrating workpiece coordinate system with edge finder.

(3) Master the method for setting Z-direction tool compensation.

1. Use of Tool Bank of Machining Center

The function of tool bank of machining center is to install all kinds of cutting tools on the machine, and you can quickly change tool from tool bank during the processing. Programming is to control the cutting tools in the machining process, and to select the suitable cutting tools in the tool bank for machining in terms of the material and shape characteristics of the parts.

The commonly-used tool bank of the machining center is divided into two categories. One is the arm-type tool bank, also known as the disc tool bank, and the other is the hat-type tool bank. In the daily use of the machining center, many factors may lead to tool disorder in tool bank. In normal processing, the corresponding tool is transferred from the tool bank of the machining center and installed on the spindle under program command to process the parts. At this time, if the machine tool is not loaded with the correct cutting tool as programmed in the previous design, or if the tool holder in the tool bank is empty, it is regarded as tool disorder in the machining center. The disorder of the machine tool directly results in the scrap of the parts, or cutting tools or machine tool strike.

The mechanical tool changer is hydraulically driven, and the position is detected by the proximity switch. In the process of tool changing, actions such as hand gripping, cutter picking, cutter changing, and cutter inserting are driven by different electromagnetic valves. There are many transmission mechanisms in this method, which are easily affected by external factors such as the stability of the hydraulic system or the sensitivity of the detection element. It is often used in the large horizontal machining center that does not require frequent change of tools.

The mechanical tool changer is driven by the AC asynchronous motor to rotate the cam mechanism. During the entire tool change process, the tool changer grabs the tool from the tool bank and places it on the spindle of the machine tool. The tool change is consistent and stable, and the time is generally about 2 seconds. However, the accuracy of the position of the spindle in loosening and gripping the tool should be in relatively high standard, and it is often used in small-sized and medium-sized machining centers that require frequent change of tools. The manipulator is driven by a servo motor, which is high in positioning accuracy, fast and stable in tool change, but is high in cost.

The instructions for tool change in the machining center are as follows.

G91G28Z0;

M6T1.

It should be noted that before changing the tool, the spindle of machine tool must be in a safe working surface and in a non-rotating state.

2. Use of the Edge Finder and the Setting of Workpiece Coordinate System

The edge finder is a detection tool used to accurately determine the center position of the workpiece in CNC machining. If the edge finder is used, the rotation speed of the spindle is 400–600 r/min. If the sample column of the edge finder is in contact with the part, due to the centrifugal force, slight contact causes obvious eccentricity for error detecting, so it is convenient to use. The accuracy of the edge finder is 0.01–0.03 mm. If the operator is highly skilled, the accuracy can reach 0.005–0.01 mm.

The edge finder is mainly used to establish the coordinate system of the workpiece in X/Y direction and the operation steps are as follows.

(1) The spindle rotates at 500 r/min. As shown in Figure 3-4-1, use the mouse to direct the mode selection knob to "MDI" (P1 in Figure 3-4-1), and press the "PROG" button (P2 in the Figure 3-4-1) in the system panel.

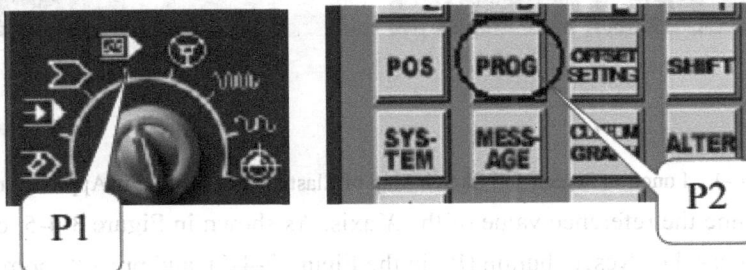

Figure 3-4-1　Entry of MDI Mode

In the order of the icons for command "S500M3" shown in Figure 3-4-2, press the corresponding buttons in the system panel and input the command "S500M3".

Figure 3-4-2　Order of the Icons for Command "S500M3"

The results will display on the CRT screen as in Figure 3-4-3 (P1 in the Figure 3-4-3). Press the "Cycle Start" button (P2 in the Figure 3-4-3). The content inputting of program on the CRT screen ends, then the program starts.

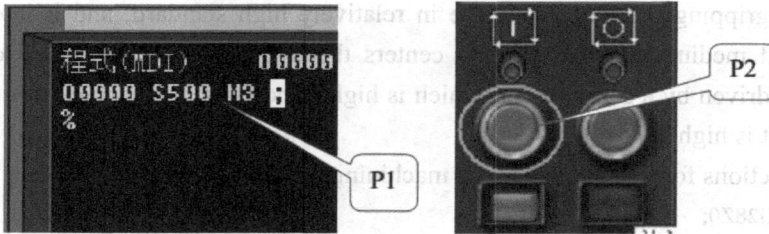

Figure 3-4-3　Input and Startup of Program

(2) Make the elastic sample pillar approach the part blank quickly, its sketch map is shown as in Figure 3-4-4(b). Use the mouse to direct the mode selection knob to "Quick Maneuver" (P1 in the Figure 3-4-4(a)), select the fast-moving axis (P2, P3, P4 in the Figure 3-4-4(a)), until the elastic sample pillar is close to the part (P5 in the Figure 3-4-4(b)).

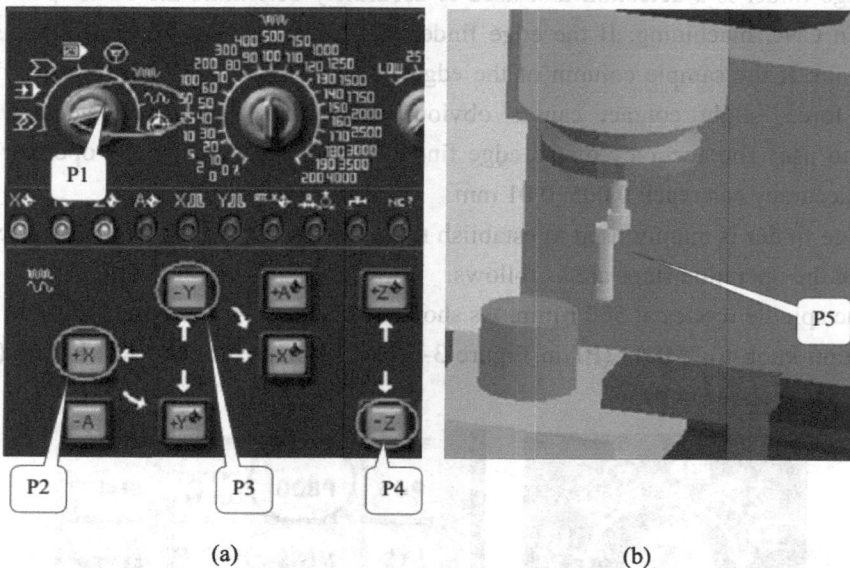

(a)　　　　　　　　　　　　　　　　　(b)

Figure 3-4-4　Function Screen and Sketch Map of Elastic Sample Pillar in Approaching Parts

(3) Determine the reference value of the X axis. As shown in Figure 3-4-5, click "RESET" in menu bar or press the "Reset" button (P1 in the Figure 3-4-5), and press "Zoom in" button (P2 in the Figure 3-4-5).

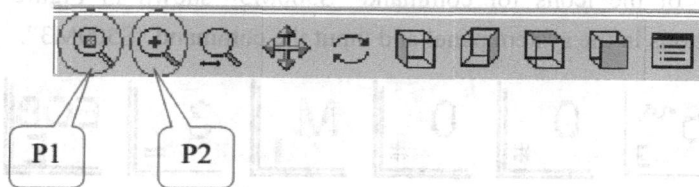

Figure 3-4-5　Adjusting Viewing Angle to Observe Parts

After the elastic sample column goes close to part, in order to ensure safe operation, you must use handwheel. Click the "HAND" button (P1 in Figure 3-4-6), and the handwheel panel appears. Adjust the handwheel to the X direction (P2 in the Figure 3-4-6), and adjust the magnifying power of moving speed (P3 in the Figure 3-4-6). Rotate the HANDLE (P4 in the Figure 3-4-6) and operate as follows, keep mouse on the handwheel, press the left button of mouse to turn the handwheel to the left, and press the right button of the mouse to turn the handwheel to the right.

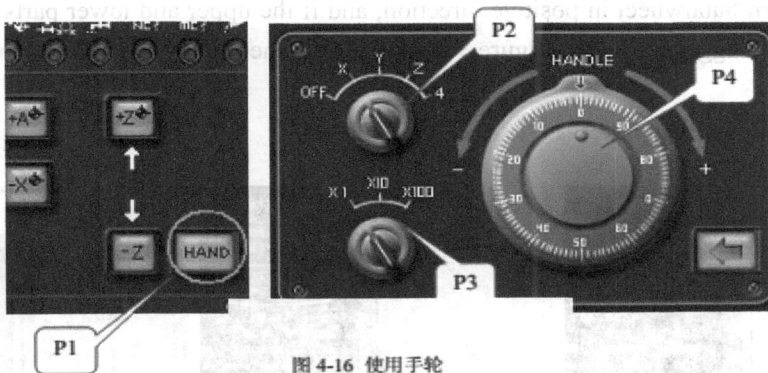

图 4-16 使用手轮

Figure 3-4-6   Use of Handwheel

Turn the handwheel to make the elastic sample pillar gradually approach the part from the right side of the part. At this time, the upper and lower parts of the sample pillar are separated, as shown in Figure 3-4-7 (P1 in the Figure 3-4-7). Turn the handwheel in negative direction, the elastic sample pillar gradually gets close to the part. In this process, you need to reduce the magnifying power of moving speed from $\times 100 \rightarrow \times 10 \rightarrow \times 1$, and reduce the eccentric distance of the upper and lower parts of the elastic sample pillar to integrate them into one (P2 in the Figure 3-4-7), the machine tool's "mechanical coordinate" X is 334.911. Continue to turn the handwheel to the left, the upper and lower parts of the sample pillar suddenly separate (P3 in the Figure 3-4-7). At this time, the machine tool's "mechanical coordinate" X is 334.910.

Note down $X_{right} = 334.911$.

Note: This value is different for different parts or locations of the parts.

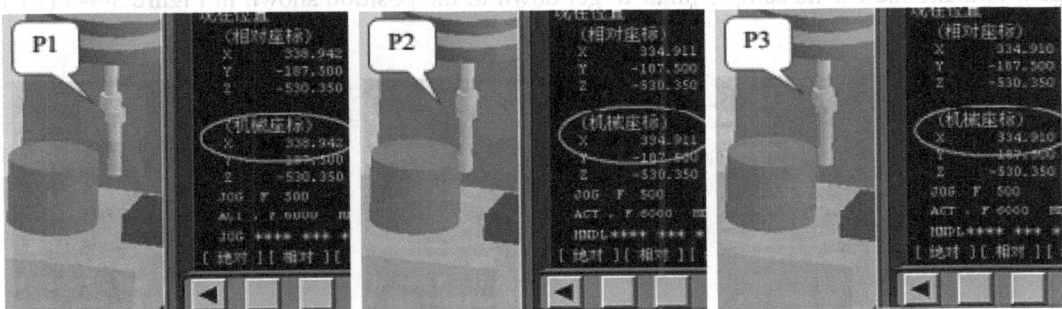

Figure 3-4-7   Determination of Reference Value on Right Side of X Axis

Remember not to move the Y axis at this time, and only the X axis and Z axis can be moved. Firstly, move the Z axis to raise the elastic sample pillar to a safe height above the part, and then move the X axis to move the elastic sample pillar to the left of the part, as shown in Figure 3-4-8. Adjust the movement speed override of the handwheel from ×100 → ×10 → ×1, make the elastic sample pillar gradually get close to the part from the left side of the part. At this time, the upper and lower parts of the sample pillar are staggered (P1 in the Figure 3-4-8). Turn the handwheel in the positive direction, and if the upper and lower parts of the sample pillar are integrated (P2 in the Figure 3-4-8), the machine tool's "mechanical coordinate" X is 265.089. Continue to turn handwheel in positive direction, and if the upper and lower parts of the sample pillar suddenly stagger (P3 in the Figure 3-4-8), the machine tool's "mechanical coordinate" X is 265.090.

Note down $X_{left} = 265.089$.

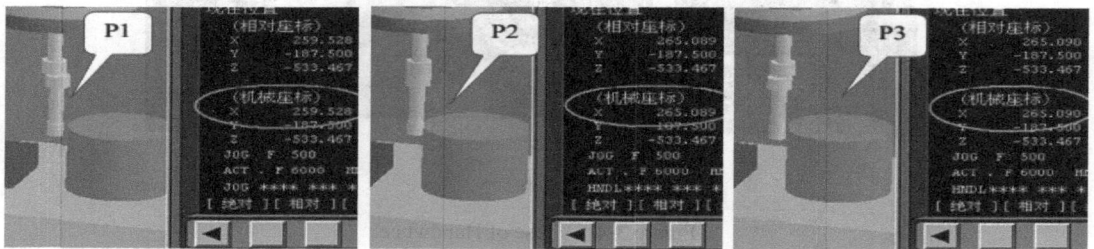

Figure 3-4-8   Determination of Reference Value on Left Side of X Axis

Below is the value of the machine tool's "mechanical coordinate" of the X axis of the part center.

$$X_{middle} = \frac{X_{left} + X_{right}}{2} = \frac{334.911 + 265.089}{2} = 300$$

(4) Determine the reference value of the Y axis. Click "View / Reset" in menu bar and press "Zoom in" button for the appropriate view. Using the handwheel, move Z axis to raise the spindle to a safe position. And move X axis for the spindle to get to the position where the mechanical coordinate value of X axis is X_{middle} (which is 300). Then move Y axis to make the spindle get to the right side of the part (one side closing to the operator). At last, move Z axis again to make the elastic sample pillar to get down to the position shown in Figure 3-4-9 (P1 in the Figure 3-4-9).

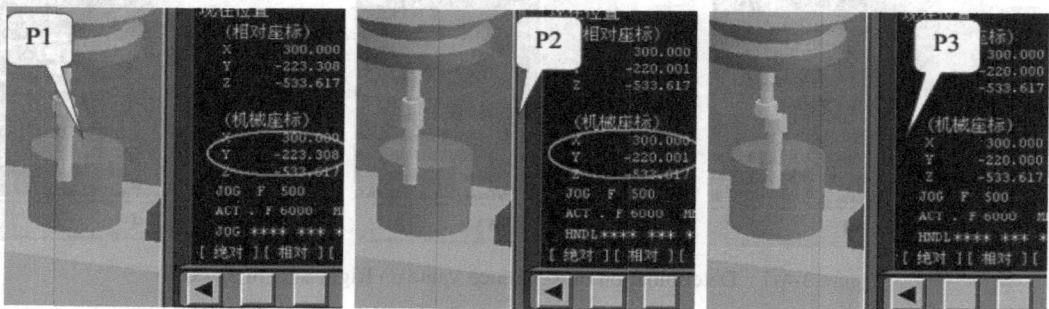

Figure 3-4-9   Determination of Reference Value on Right Side of Y Axis

Use the handwheel to move the elastic sample pillar, make it get closer the part. At this time, the upper and lower parts of the sample pillar are staggered (P1 in the Figure 3-4-9). Turn the handwheel in positive direction, adjust the magnifying power of moving speed from ×100 → ×10 → ×1. When the upper and lower parts of the sample pillar are integrated (P2 in the Figure 3-4-9), the machine tool's "mechanical coordinate" Y is −220.001. Continue to turn the handwheel in the positive direction, the upper and lower parts of the sample pillar suddenly stagger (P3 in the Figure 3-4-9), and at this time the "mechanical coordinate" Y is −220.000.

Note down Y_{right} =−220.001.

It should be noted that at this time, the mechanical coordinates of the X axis should be kept at 300. Do not move the X axis and move only the Y axis and Z axis. Firstly, move the Z axis to raise the elastic sample pillar to a safe height above the part, and then move the Y axis to move the elastic sample pillar to the left side of the part (one side away from the operator), as shown in Figure 3-4-10. Adjust the magnifying power of movement speed of the handwheel from ×100 → ×10 → ×1, so that the elastic sample pillar gradually approaches the part from the left side of the part. At this time, the upper and lower parts of the sample pillar stagger (P1 in the Figure 3-4-10). Turn the handwheel in positive direction, and when the upper and lower parts of the sample pillar are integrated (P2 in the Figure 3-4-10), the machine tool's "mechanical coordinate" Y is −149.999. Continue to turn the handwheel in negative direction, and when the upper and lower parts of the sample column suddenly stagger (P3 in the Figure 3-4-10), the machine tool's "mechanical coordinate" Y is −150.000.

Note down Y_{left}=−149.999 and

$Y_{middle} = (Y_{left} + Y_{right})/2 = (−220.001 − 149.999)/2 = −185.0$

The machine tool's "mechanical coordinate" value of the Y axis of the part center is $Y_{middle} = (Y_{left} + Y_{right}) / 2 = (−220.001−149.999) / 2 = −185.0$.

Up to now, the machine tool's "mechanical coordinate" value of the X axis and Y axis of the part center has been known, which is (300, −185). This value is to be put into the user's coordinate system.

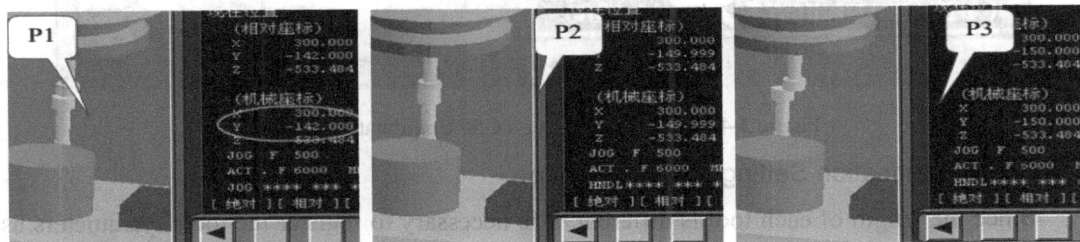

Figure 3-4-10　Determination of Reference Value on Left Side of Y Axis

After above operation, raise the elastic sample pillar to a safe height above the part. Press "RESET" to stop the spindle rotation. Then hide the handwheel (shown as P1 in Figure 3-4-11) and remove the reference tool (shown as P2 in Figure 3-4-11).

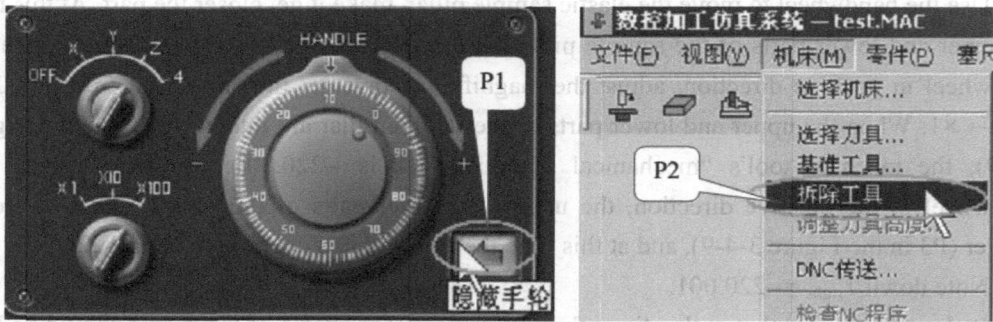

Figure 3-4-11 Hide Handwheel and Remove Reference Tool

The determination of reference value of the Z axis is also related to the actual tool used.

(5) Establish the user's coordinate system (G54). As shown in Figure 3-4-12, press "OFFSET SETTING" (P1 in the Figure 3-4-12) with the mouse, and then press the corresponding "soft key" button (P2 in the Figure 3-4-12) in the CRT to enter the user's coordinate system. Since the commonly used coordinate system is G54, move the cursor (P3 in the Figure 3-4-12) and input the machine tool's "machine coordinate" values of the X axis and Y axis of the part center into the G54 coordinate system (P4 in the Figure 3-4-12). Note that the value of the Z axis in the G54 coordinate system should be kept at zero.

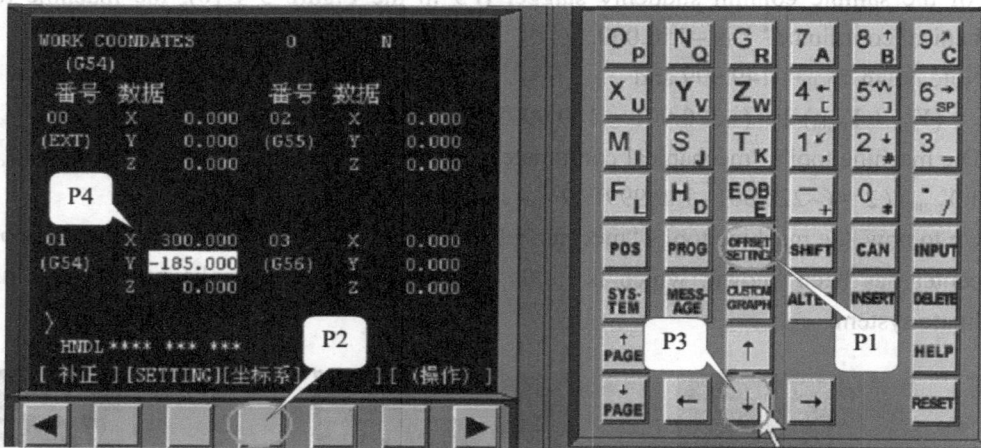

Figure 3-4-12 Establish User's G54 Coordinate System

3. Compensation Setting for Tool's Shape

Since the length of each tool is different, it is necessary to register the tools' parameters as follows.

(1) Change the tool No. 1, and input a program to equip the spindle with the T1Φ12 end milling cutter. As shown in Figure 3-4-13, direct the mode selection knob for "Edit" (P1 in the Figure 3-4-13) with the mouse, and press the "PROG" button (P2 in the Figure 3-4-13) in the system panel.

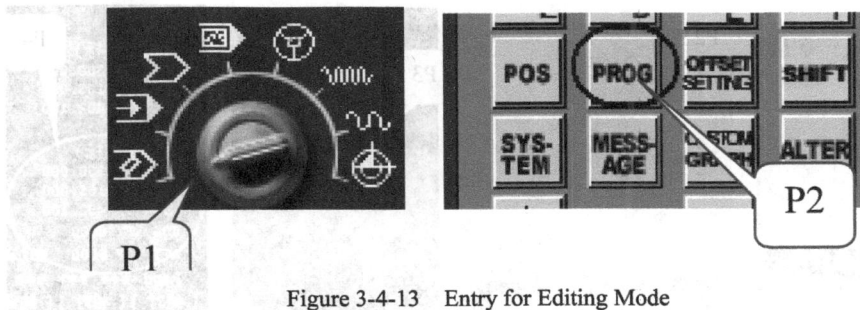

Figure 3-4-13　Entry for Editing Mode

Press the corresponding button in the system panel to input the command.

O00002;

G91G28Z0;

M6T1;

G54G90G00X0Y0.

After above operation, press "RESET" key to return to the head of program, and the program input is completed.

As shown in Figure 3-4-14, direct the mode selection knob to "Automatic Processing" (P1 in the Figure 3-4-14) with the mouse. The result to be displayed on the CRT screen is shown (P2 in the Figure 3-4-14). Then press the program "Cycle Start" button (P3 in the Figure 3-4-14).

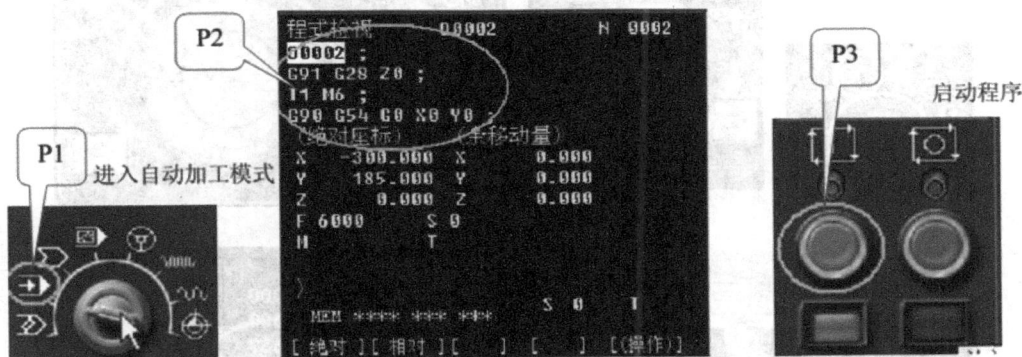

Figure 3-4-14　Startup of Tool Change Program

The machine tool automatically equips the spindle with the tool No. 1. After the tool change is completed, the spindle automatically moves above the workpiece. In manual mode, let the spindle of the machine tool approach the workpiece quickly. When it is about 50 mm away from the upper surface of the workpiece, change the working mode and use the handwheel to shake the Z axis of the machine tool, and let the spindle slowly approach the workpiece.

(2) Determine the length compensation value of tool No. 1 (H1). As shown in Figure 3-4-15, click the menu "Feeler gauge check / 100 mm (Gauge Block)" (P1 in the Figure 3-4-15). At this time, the display of the machine tool is divided into two parts (P2 and P3 in the Figure 3-4-15), and a message box of the feeler gauge check appears. Click on the system panel with the mouse, and then click the "General" soft key at the bottom of the CRT screen to display the machine coordinate system (P4 in the Figure 3-4-15).

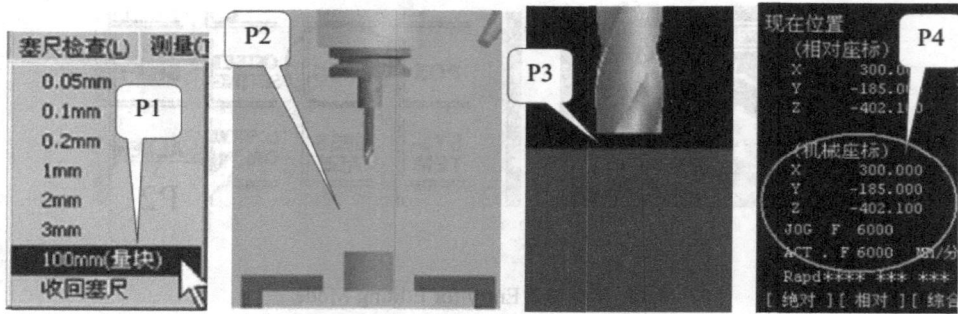

Figure 3-4-15   Z-Direction Tool Setting

As shown in Figure 3-4-16, direct the mode selection knob to "Handwheel" (P1 in the Figure 3-4-16) with the mouse, and click the "HAND" icon (P2 in the Figure 3-4-16) to display the HANDLE (P3 in the Figure 3-4-16). Adjust the axial movement to Z axis (P4 in the Figure 3-4-16), use the left button of the mouse to click the handwheel, and let the tool gradually approach the part from the top. In this process, adjust the magnifying power of movement speed from large to small, that is, ×100 → ×10 → ×1 (P5 in the Figure 3-4-16). The message box of the feeler gauge check now shows whether the Z-direction of the tool distant from the part.

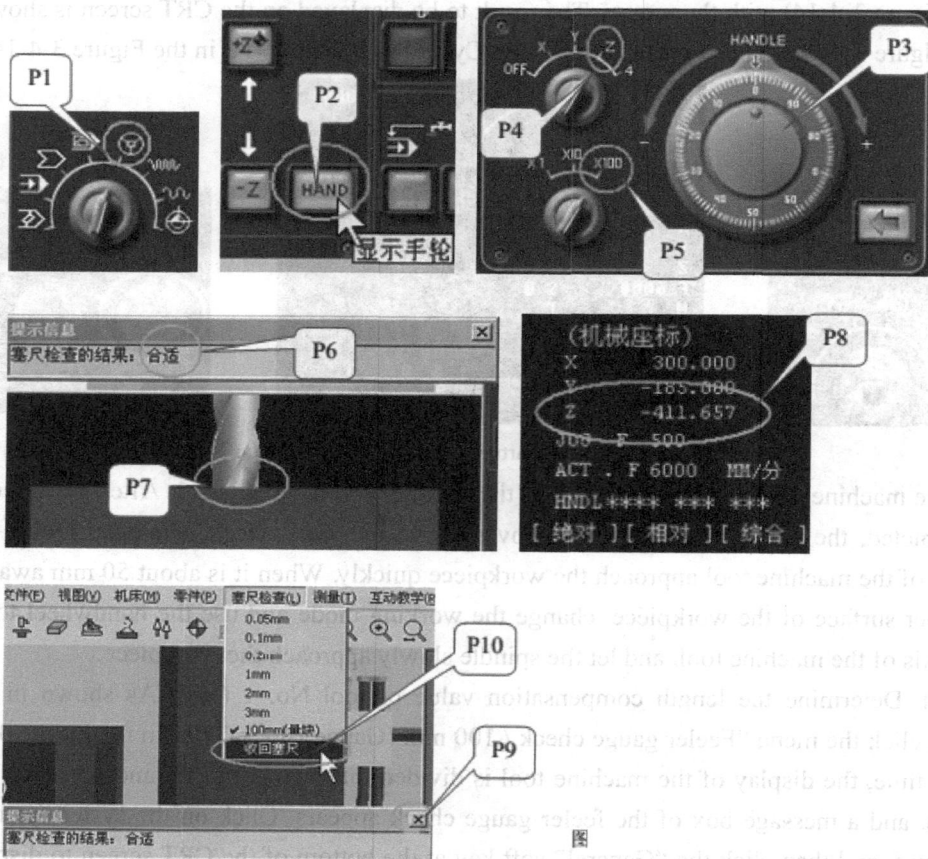

Figure 3-4-16   Determine Z-direction Dimensions of Tool

If the result of feeler gauge check is "appropriate" (P6 in the Figure 3-4-16), it indicates that the distance of the tool from the gauge block (P7 in the Figure 3-4-16) is 100 mm.

At this time, the machine coordinate system "mechanical coordinate" Z is -411.657 (P8 in the Figure 3-4-16).

As shown in Figure 3-4-16, close the message box of the feeler gauge check (P9 in the Figure 3-4-16), and click the "feeler gauge check / retract feeler gauge" (P10 in the Figure 3-4-16) in the menubar. If the machine tool is a CNC milling machine and if only one tool is used to process parts, the $H1$ of the tool length compensation value is as follows.

$$H1 = Z1 - Z_{\text{gauge block}} = -411.657 - 100.0 = -511.657$$

The machine tool used now is the machining center. With several milling cutters, the previous process needs to be repeated several times.

(3) Register the compensation value of the tool. Register the H value of each tool for the tool length compensation, and the operation steps are shown in Figure 3-4-17.

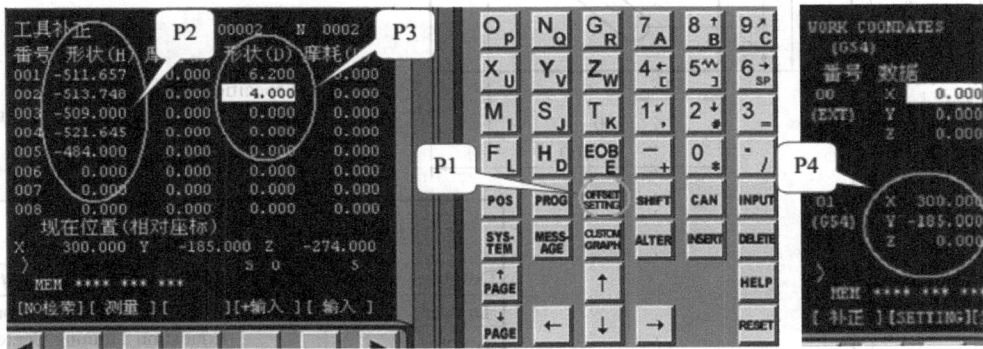

Figure 3-4-17　Register Compensation Value of Tool Length

Press the "OFFSET SETTING" button (P1 in the Figure 3-4-17) with the mouse, and the tool compensation screen (P2 in the Figure 3-4-17) is displayed on the CRT interface. Under the shape (H) item, input the measured $H1-H5$ into the screen sequentially (P2 in the Figure 3-4-17). As for the input method, please refer to the input value of G54 coordinate system in Figure 3-4-12.

The length compensation values are to be requested in the program with the G43HXX command. If there is no G43 command in the program, these length compensation values are invalid. In the program, $D1$ (6.2) and $D2$ (4.0) are needed and these two tool radius programming values are input to the shape (D) item (P3 in the Figure 3-4-17). In the program, use the G41DXX command to request these compensation values. If there is no G41 command in the program, these radius compensation values are invalid.

Note that using the method for tool setting, the Z axis value in G54 must be kept at 0 (as P4 in the Figure 3-4-17).

4. Skills Training

Complete the task sheet as Table 3-4-1 based on what you have learned above.

Table 3-4-1　Task sheet

Task name	Workpiece calibration for simple parts in the machining center program		
Class		Group No.	
Task assignment			
Workpiece for calibration	 Technical specifications : 1. Sharp-corner bevel edge, seamed-corner bevel edge C2. 2. Arrange annealing for the part.		
Methods and steps for calibration			

Within the workpiece drawing:

Dimensions: 2, 15, 10, 80, 60, 15, 30, 80, 100

Part Name		Proportion	01
		Quantity	
Designer		Weight	45#
Drawer		Lancang-Mekong Vocational Education Training Center	
Auditor			

Task 5   Debugging and Operation of CNC

Machining Program

【Knowledge objectives】

(1) Understand the operation of the machining center in automatic mode.

(2) Learn to run the machining program on a trail basis and debug the processing program.

【Ability objectives】

(1) Master the method for operation in automatic mode.

(2) Master the methods of test running and program debugging.

1. Operation of the Machining Center in Automatic Mode

1) Automatic / continuous mode

① Check whether the machine tool returns to zero. If not, first return the machine to zero.

② Use the CNC program or write a program by yourself.

③ Check whether the mode selection knob on the control panel is set to the AUTO position. If not, click the mode selection knob by mouse for AUTO position and enter the automatic processing mode.

④ Press the "Start" button in [Start Hold Stop], and the CNC program starts to run.

The operation can be discontinued.

The CNC program can be in pause, stop, emergency stop and re-run operation as needed during the running process.

When the CNC program is in running, click the "Hold" button in [Start Hold Stop], and the program is paused. Click the "Start" button again, and the program continues to run from the paused line.

When the CNC program is running, click the "Stop" button in [Start Hold Stop], the program stops running. Click the "Start" button again, the program runs again from the beginning.

When the CNC program is running, press the "Emergency Stop" button [⟳], and the CNC program is interrupted. To continue the operation, release the emergency stop button first, and then press the "Start" button in [Start Hold Stop], and the remaining CNC program to be executed will run as an independent program starting from the interrupted line.

2) Automatic / single block mode

① Check whether the machine tool returns to zero. If not, first return the machine tool to zero.

② Use the CNC program or write a program by yourself.

③ Check whether the mode selection knob on the control panel is set to the AUTO position. If not, click the mode selection knob with the left or right button of the mouse for AUTO position and enter the automatic processing mode.

④ Set the "Single Block" switch to "on".

⑤ Press the "Start" button in , and the CNC program starts to run.

Note that when executing each line of the program in automatic / single block mode, you need to click the "Start" button in once.

Set the "Opt Skip" switch to "on" and the skip symbol "/" in CNC program is valid. Set the "M01 Stop" switch to "on" and the "M01" code is valid.

Adjust the feed speed (F) adjustment knob as required to control the feed speed of operating the CNC program, and the adjustment range is 0%–150%. If the mode selection knob on the control panel is switched to DRY RUN at this time, it means that the feed is at G00 speed.

Press the key RESET to reset the program.

2. "Running on a trial basis" and Debugging of Processing Program

Having introduced the CNC program, the running track can be checked.

Switch the mode selection knob of the operation panel to AUTO or DRY RUN, click the AUX GRAPH in the control panel, and switch to the checking the running track mode. Then click "Start" in on the operation panel to observe the running track of the CNC program. It is also possible to make an all-round dynamic observation of the three-dimensional running track by dynamic rotation, dynamic zooming, dynamic translation and other methods in the "View" menu.

Note that pause operation, stop operation, single block execution, etc. are also effective when in checking the running track mode.

3. Skills Training

Complete the task sheet as Table 3-5-1 based on what you have learned above.

Table 3-5-1　Task sheet

Task name	Methods to test the running operation for simple parts	
Class		Group No.
Task assignment		
Precautions for the automatic mode		
Precautions for running test		

Module 4

Basic Preparation and Inspection of Machining Center

Task 1   Preparation on Operation

【Knowledge objectives】

(1) Understand the characteristics of process planning in plain milling of parts.

(2) Understand the characteristics of process programming in plain milling of parts.

【Ability objectives】

(1) Master the machining process of plain milling.

(2) Master the methods for plain milling programming.

1. Process Planning in Plain Milling of Parts

In plain milling, the planar parts are usually cut by the side edge of an end milling cutter. Due to the different movement path and direction of the cutter, it may perform down milling or reverse milling, and the surface quality of the parts obtained by different machining routes is also different.

Looking at the feed direction of the tool, if the workpiece is located to the right side of the feed direction of the milling cutter, the feed direction is called clockwise. Conversely, when the workpiece is located to the left, the feed direction is defined as counterclockwise. If the rotation direction of the milling cutter is opposite to the feed direction of the workpiece, it is called reverse milling, as shown in Figure 4-1-1; if the rotation direction of the milling cutter is the same as the workpiece feed direction, it is called down milling, as shown in Figure 4-1-2.

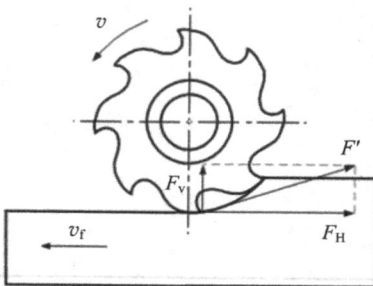

Figure 4-1-1   Reverse Milling

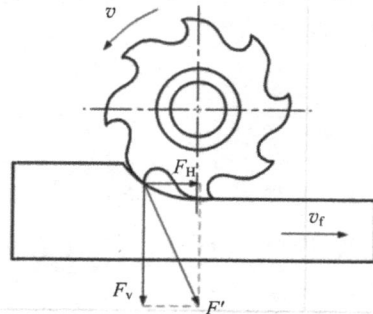

Figure 4-1-2   Down Milling

During reverse milling, the cutting thickness of each cutter changes gradually from small to large, and the cutter teeth cut in from the side of processed surface, which is beneficial to the use of milling cutter. However, as the cutter teeth of the milling cutter cannot cut into the metal layer immediately after contacting the workpiece, they slide on the surface of the workpiece for a short distance. During the sliding process, a large amount of heat is generated due to strong friction. At the same time, a hardened layer is easy to form on the surface to be machined, which reduces the durability of the tool, affects the surface roughness of the workpiece, and brings disadvantages to cutting.

During down milling, the cutting thickness changes gradually from large to small. The moment the cutter teeth start to contact the workpiece, the cutting thickness is the largest, and the cutting starts from the hard layer on the surface of the workpiece in high machining efficiency, at this time, the cutter teeth are subject to a great impact load, and the milling cutter becomes blunt quickly, but there is no slippage in the cutting process. The power consumption of down milling is smaller than that of reverse milling. Under the same cutting conditions, the power consumption of down milling is 5%–15% lower than that of the reverse milling (the power consumption can be reduced by 5% when in milling carbon steel and 14% when in milling difficult-to-machine materials). At the same time, down milling is also more conducive to chip removal. However, if the direction of horizontal milling force is consistent with the workpiece feed movement direction, and if the force of cutter teeth on workpiece is large, considering the gap between the lead screw and the nut of the worktable, the worktable waggles, which not only destroys the smoothness of the cutting process and affects the machining quality, but also damages the tool in severe cases.

At present, CNC machine tools usually have a clearance eliminating mechanism, which can reliably eliminate the gap between the table feed screw and the nut, and prevent vibration during the milling process. Therefore, if the workpiece has no hard layers and the process system is rigid enough, the CNC milling should adopt the down milling as much as possible to reduce the roughness of the surface of the machined part and ensure the dimensional accuracy. However, if there are hard layers and slags on the cutting surface and there is significant unevenness on the surface of the workpiece, such as forging blank, the reverse milling should be used.

In milling plane parts, the feed mode before cutting must also be considered. There are two forms of feed mode before cutting, vertical feed mode and horizontal feed mode.

As shown in Figure 4-1-3 (a), in milling the contour of the outer surface, in order to reduce the marks and ensure the surface quality, the milling cutter should cut along the tangent extension line of a point on the contour curve of the part. If the cut-in and cut-out distances are limited, straight milling feed followed by arc milling feed as a transition can be used, as shown in Figure 4-1-3 (b). The same processing method can be adopted to mill the inner contour surface, as shown in Figure 4-1-4.

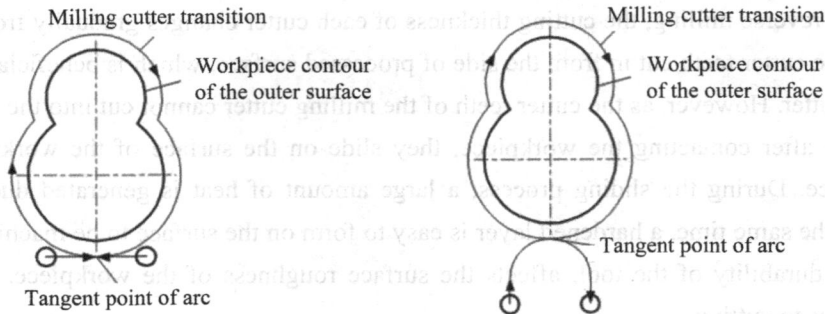

Figure 4-1-3　Milling Contour of Outer Surface

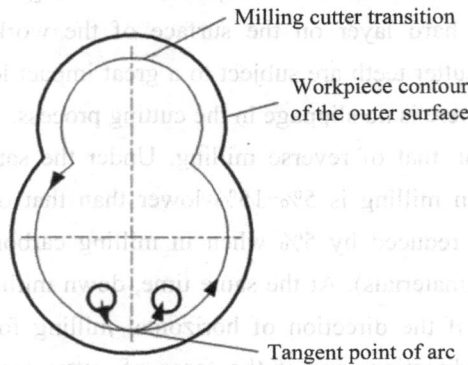

Figure 4-1-4　Milling Contour of Inner Surface

2. Process Programming in Plain Milling of Parts

1) Quick point positioning (G00)

With G00 command of quick point positioning, the tool is instructed to move in the point control mode from the point where the tool is located to the target point at the fastest speed.

The program format for three-axis linkage is G00X_Y_Z_.

X_Y_Z_ is the coordinate value of the target point. If absolute value is used for programming, X, Y and Z are the coordinate values of the target point in the workpiece coordinate system; as for programming with incremental values, X, Y and Z are the incremental coordinate values of the target point relative to the starting point. The rapid traverse speed of G00 is set by the machine tool manufacturer for each axis separately, and each axis moves quickly and independently at the set speed. The tool trajectory during positioning is jointly determined by the rapid movement speed of each axis and it cannot be ensured that each axis reaches the end point at the same time. Therefore, the combined trajectory of each axis linkage is not necessarily a straight line. The rapid traverse speed of G00 cannot be changed by program instructions, but it can be changed by rapid traverse adjustment knob on the control panel. In G00 positioning mode, the tool starts to accelerate from the starting point to the predetermined speed, decelerates before reaching the end point, and stops by precise positioning. G00 is used only for rapid positioning, not for cutting.

2) Linear interpolation (G01)

The tool feeds in the command of linear interpolation at the speed specified in this block for machining linear paths, as shown in Figure 4-1-5.

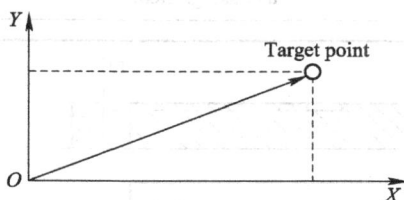

Figure 4-1-5　Linear Interpolation

The program format for three-axis linkage is G01X_YZ_F_.

X_Y_Z_ is the coordinate value of the target point, F_ is the feed rate, and the actual feed rate of each axis is the projection component of F on each axis.

As shown in Figure 4-1-6, assuming that the starting position of the tool is point A, the contour program is compiled along the path of $A \rightarrow B \rightarrow C \rightarrow D \rightarrow E$.

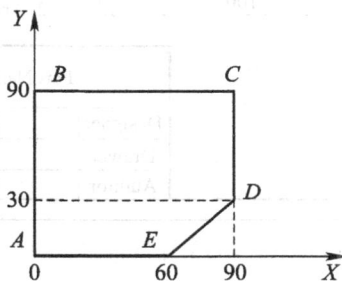

Figure 4-1-6　Linear Contour Programming

Below is absolute programming.

 N1G90G01X0Y90.0F100; A→B

 N2G01X90.0; B→C

 N3Y30.0; C→D

 N4X60.0Y0; D→E

 N5X0; E→A

Below is incremental programming.

 N1G91G01X0Y90.0F100; A→B

 N2G01X90.0; B→C

 N3Y-60.0; C→D

 N4X-30.0Y-30.0; D→E

 N5X-60.0; E→A

3. Skills Training

Complete the task sheet as Table 4-1-1 based on what you have learned above.

Table 4-1-1　Task sheet

Task name	CNC machining of flat parts		
Class		Group No.	
Task assignment			
Drawings for processing	 Technical specifications : 1. Sharp-corner bevel edge, seamed-corner bevel edge C2. 2. Arrange annealing for the part.		
	Part Name	Proportion	01
		Quantity	
	Designer	Weight	45#
	Drawer	Lancang-Mekong Vocational Education Training Center	
	Auditor		
Programming the processing			

Task 2　Dimensional Accuracy Control in Plain Milling

【Knowledge objectives】

(1) Learn the methods for automatic processing of machining center.

(2) Understand the methods for controlling dimensional accuracy in machining.

(3) Learn to check the accuracy of plain milling of parts.

【Ability objectives】

(1) Master the methods for automatic processing of machining center.

(2) Master the methods for controlling dimensional accuracy in machining.

(3) Master the methods for inspecting the processing accuracy.

1. Automatic Processing of Machining Center

1) Technical requirements for the surface being machined

The technical requirements for the surface being machined are mainly on its flatness and surface roughness. And for some parts, there may be other requirements for physical properties.

(1) Flatness. Flatness refers to the flatness of the surface of a workpiece. As shown in Figure 4-2-1, $\boxed{\square \mid 0.025(-)}$ means that the flatness error of the top surface of the workpiece is not allowed to exceed 0.025 mm in the entire surface, that is, the flatness tolerance is 0.025 mm. The symbol (−) indicates that the surface is allowed only to be concave.

The rectangular workpiece shown in Figure 4-2-1 has a total of 6 surfaces. It is required to ensure the perpendicularity between the two adjacent surfaces (A and B), the parallelism between the two opposite surfaces (A and D), and the dimensional accuracy between two opposite surfaces (A and D, and B and C).

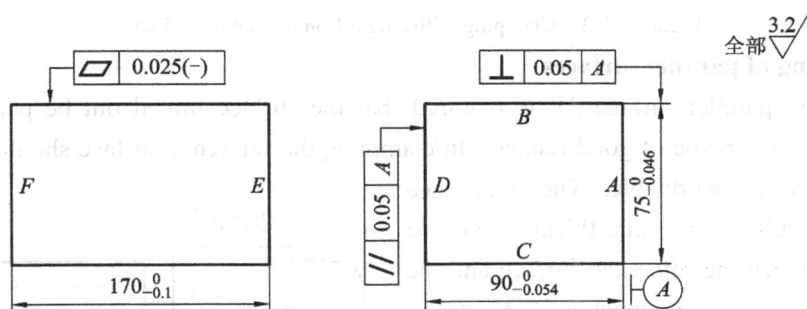

Figure 4-2-1　Rectangular Workpiece

(2) Surface roughness. Surface roughness refers to the microscopic geometrical feature shown by a smaller distance of the processed surface and peak-valley. The symbol $\overset{3.2}{\bigvee}$ in Figure 4-2-1 shows that the surface roughness of the processed surface Ra is 3.2 μm.

2) Milling of vertical surfaces

(1) Clamping a workpiece on the machine flat vise for processing. Ensure that the reference surface of the workpiece is firmly attached to the fixed jaw as shown in Figure 4-2-2(a). Clean the fixed jaw and the reference surface, and place a round rod at the movable jaw, as shown in Figure 4-2-2 (b). If there is no round rod, a long narrow piece of thick copper sheet can also be placed.

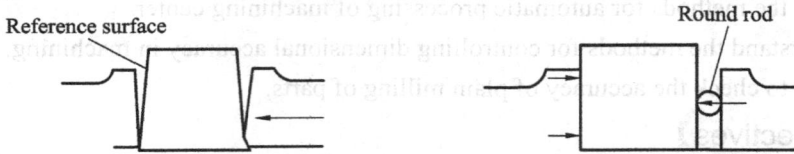

Reference surface   Round rod

(a) Reference surface attached to the fixed jaw   (b) Clamping with round rod on the movable jaw

Figure 4-2-2   Clamping by Machine Flat Vise

(2) Clamping the workpiece with angle iron or with clamping plate for processing. As shown in Figure 4-2-3 (a), the workpiece is clamped with angle iron and processed on a horizontal milling machine. Figure 4-2-3(b) shows that the workpiece is clamped with clamping plate and processed on a vertical milling machine. This processing method is suitable for workpieces with relatively wide reference surface and narrow machining surface. By this method, the factors that affect the verticality are the cylindricity and the reciprocation of the milling cutter.

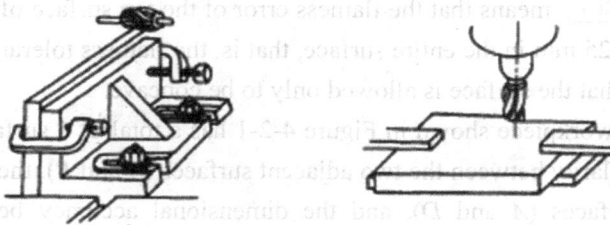

(a) Clamping with angle iron   (b) Clamping with clamping plate

Figure 4-2-3   Clamping with Angle Iron or Clamping Plate

3) Milling of parallel surfaces

In milling parallel surfaces, it is required that the surface milled out be parallel to the reference surface, and be of good flatness. In clamping, the reference surface should be parallel to the surface of worktable. Therefore, two parallel iron pads of the same thickness should be placed between the reference surface and the surface of lead rail of the machine flat vise, as shown in Figure 4-2-4. Even for thick workpieces, it is better to pad two thin copper sheets of equal thickness in order to check whether the reference surface is parallel to the lead rail of machine flat vise.

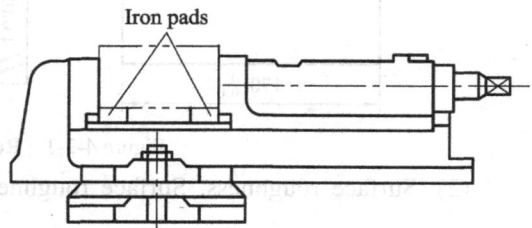

Iron pads

Figure 4-2-4   Parallel Iron Pads

If there are steps on the workpiece in milling the parallel surface on a vertical milling machine, the workpiece can be clamped directly on the surface of the worktable with a clamping plate for the reference surface to fit the worktable's surface, and then the parallel surface can be milled with a face milling cutter, as shown in Figure 4-2-5(a). For large-sized workpieces with out side steps, its parallel surfaces can be milled with a face milling cutter on a horizontal milling machine, as shown in Figure 4-2-5(b). In clamping, the positioning key can be used to make that the reference surface is parallel to the longitudinal direction. If the bottom surface is perpendicular to the reference surface, no further correction is required. If not, you need to adjust it with iron pads or re-mill the bottom surface. Having adjusted, you need to check the reference surface with a 90° square ruler. To meet the requirement for high accuracy, the dial indicator can be fixed on the suspension beam through the scale frame, the workbench can be moved up and down, and then the reference surface can be corrected.

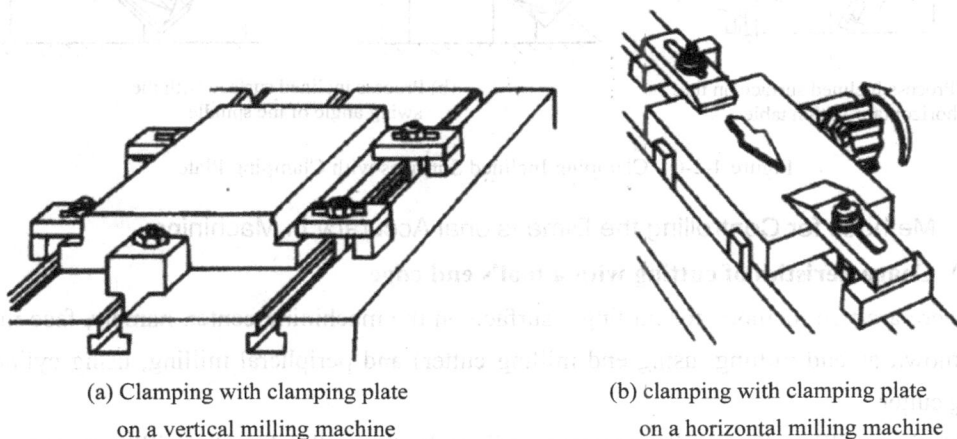

(a) Clamping with clamping plate　　　　　　(b) clamping with clamping plate
on a vertical milling machine　　　　　　　　　on a horizontal milling machine

Figure 4-2-5　Clamping with Clamping Plate for Parallel Surfaces

4) Size control between two parallel surfaces

In producing single pieces, the cycle of milling → measurement → milling is generally adopted until the size is accurate. It should be noted that the amount of lift or offset of the milling cutter in rough milling is not equal to that in finish milling, which must be considered in controlling the size.

To meet the requirements for high dimensional accuracy, it is necessary to conduct a semi-finish milling after rough milling, with a margin of about 0.5 mm. To measure the size of workpieces after rough milling or semi-finish milling, if conditions allowed, it is better not to remove the workpieces but to measure them on a workbench.

5) Milling of inclined surfaces

The so-called inclined surface refers to the surface of the part that is inclined to the reference surface, and they intersect to form an angle. Generally, there are two ways.

One is the inclined surface of a large slope indicated in degrees. For example, the angle between the inclined surface of the dovetail groove and the reference surface is 55°.

The other is the inclined surface of a small slope indicated by ratios. For example, the machined material is at a length of 100 mm, the difference between the two ends is 1 mm, which is represented by the slope "$\angle 1$: 100". The milling of an inclined surface is actually like milling a surface. All that needs to be done is tilt the workpiece to be machined at an angle relative to the workpiece spindle. The horizontal rotation table in Figure 4-2-6 (a) or the swing angle of the spindle in Figure 4-2-6 (b) of the machining center can often be used for machining.

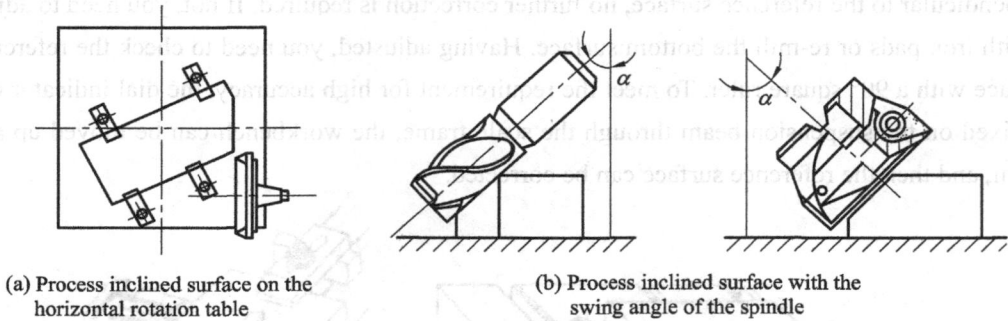

(a) Process inclined surface on the horizontal rotation table

(b) Process inclined surface with the swing angle of the spindle

Figure 4-2-6   Clamping Inclined Surfaces with Clamping Plate

2. Methods for Controlling the Dimensional Accuracy in Machining

1) Characteristics of cutting with a tool's end edge

There are two methods for milling a surface on the machining center, namely, face milling (also known as end milling, using end milling cutter) and peripheral milling, using cylindrical milling cutter.

(1) Face milling. Face milling refers to milling by the teeth edge of a milling cutter, using a cutting edge distributed on the milling cutter's end to process a surface. In milling surfaces, the face milling method is used to process the wider, larger surfaces in particular.

Face milling can be processed in two ways, symmetrical milling and asymmetrical milling (see Figure 4-2-7).

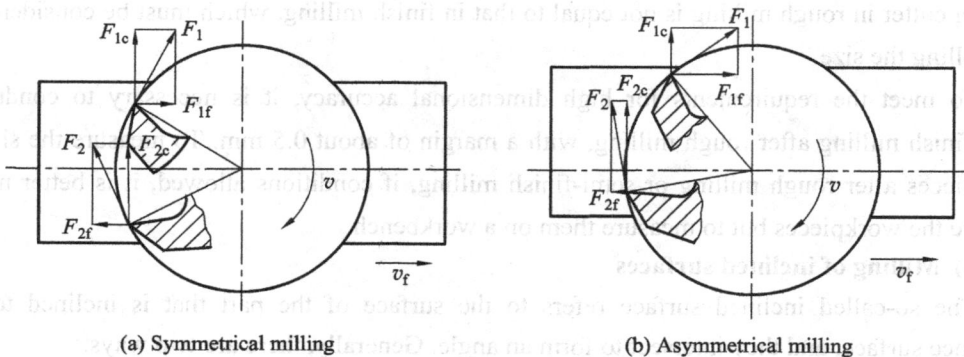

(a) Symmetrical milling

(b) Asymmetrical milling

Figure 4-2-7   Symmetrical Milling and Asymmetrical Milling

The average cutting thickness of symmetrical milling is relatively large, so this milling method is suitable for milling hardened steel and machine tool guide rail. The vibration of is relatively large by symmetrical milling, so the rigidity of machine tool and of milling system should be strong. It is not suitable for milling narrow working surfaces.

Asymmetric milling can be divided into asymmetric reverse milling and asymmetric down milling. In asymmetric reverse milling, the cutter teeth cut into the workpiece from the thin position to the thick position, so the vibration is small and the work is balanced when cutting in, and there is no sliding caused by the zero cutting thickness which happens in cutting by using the cylindrical milling cutter for reverse milling. In asymmetric down milling, the cutter teeth cut into the workpiece from the thin position to the thick position and cut out from the small cutting thickness. In machining stainless steel and other materials with large deformation coefficient and severe cold work hardening, asymmetric down milling method should be used. In asymmetric down milling, the vibration of cutter teeth in cutting into the workpiece is greater than that in asymmetric reverse milling, and the gap between the feed screw and the nut of the worktable should be eliminated for avoiding the worktable movement caused by the excessive horizontal milling force.

The flatness of the surface machined by the face milling is determined mainly by the perpendicularity of the spindle axis of the milling machine and the feed direction, as shown in Figure 4-2-8(a). If the spindle of the milling machine is perpendicular to the feed direction, the path for the tool tip to rotate is a circular ring parallel to the feed direction, which cuts out a surface. At this time, the cutting edge of the milling cutter mills out a net-like pattern on the surface of the workpiece. If the spindle of the milling machine is not perpendicular to the feed direction, the machining work is equivalent to cutting workpiece's surface into a concave surface with a titled ring. At this time, a one-way curved pattern is formed on the surface of the workpiece. In the milling process, if the feed direction is from the higher side of the tool tip to its lower side, a dragging occurs; otherwise, there is no dragging. In actual work, even if the spindle of the milling machine is perpendicular to the feed direction, due to the gap between the spindle bearing as well as the poor rigidity of the fixture and tool, the dragging also occurs, which affects the roughness of the workpiece's surface. In order to avoid the dragging for above reasons, the spindle of the milling machine can be made slightly non-perpendicular to the feed direction, and the feed direction can be changed to be from the lower side of the tool tip to the higher side. The amount of non-perpendicularity should be controlled under the limit of flatness tolerance.

(2) Peripheral milling. Peripheral milling refers to the milling by the peripheral teeth edge of the milling cutter, using the cutting edge of the milling cutter distributed on the cylindrical surface to mill and form a surface. Peripheral milling is suitable for processing narrow surfaces. For a workpiece with surfaces of a width greater than 120 mm, it is better to use face milling except in special cases. As shown in Figure 4-2-8 (b), the flatness of the surface milled by the peripheral milling depends mainly on the cylindricity of the milling

cutter. In peripheral milling since the milling surface is performing linear motion under the cylinder, the cylindrical milling cutter must be in cylindrical shape to attach to the cylinder milling surface during fine milling.

The spindle of the milling machine is perpendicular to the feed direction

The spindle of the milling machine is not perpendicular to the feed direction

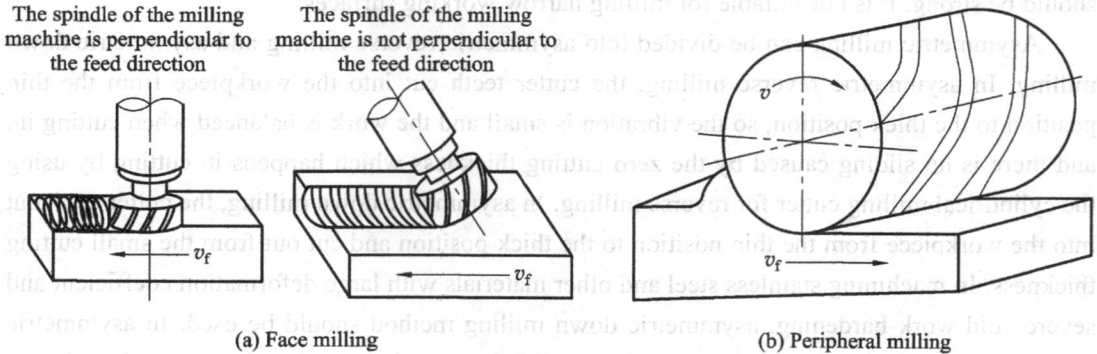

(a) Face milling

(b) Peripheral milling

Figure 4-2-8   Methods for Milling

(3) Comparison between face milling and peripheral milling.

① For the reasons that the face milling cutter bar is short and rigid, that the milling blade is easy to clamp, and especially that the milling blade is indexable, face milling is suitable for milling at a high speed and with great strength, which can significantly improve productivity and reduce surface roughness.

② The maximum diameter of the face milling cutter is about 1 m, which can mill out a wide surface at one time. At the same time, there are many cutter teeth working, so the vibration is small and work efficiency is high.

③ The requirement on sharpening of the face milling cutter is not as strict as that of the cylindrical milling cutter. For a face-milling cutter, if the edges of the teeth of each cutter are ground unevenly, and if the radius direction is different, there is no effect on the flatness of the milled plane except the negative influence on the smoothness of the part surface. The surfaces obtained by face milling can only be concave. From the perspective of the use of parts, most of them are only allowed to be concave but not convex. However, for the surfaces obtained by peripheral milling, both convex and concave may occur. Therefore, it is more reasonable to mill surfaces by the face milling method.

④ For the same amount of milling, and for when polished blades is not used on the face milling cutter, if we apply peripheral milling the surface roughness is relatively small.

2) Tool path of plain milling

For the large surface of a workpiece which cannot be cut with one feed, multiple feeds have to be used, in which case overlapping tool marks occur between two feeds. Generally, there are three feed modes for large area milling.

(1) Circular feed mode (see in Figure 4-2-9 (a)). By this machining mode there result the shortest tool stroke and the highest production efficiency. If a right-angle turn is used, the feed direction must be switched at the four corners of the workpiece, causing the tool to stop at one

position and then feed again, so that the four corners of the workpiece are cut excessively in a thin layer, which affects the flatness of the machining surface. Therefore, try to use arc milling feed as a transition at the corner.

(2) Peripheral feed mode (see Figure 4-2-9 (b)). The tool stroke by peripheral feed mode is longer than that by circular feed mode. Since the four corners of the workpiece are cut twice by the horizontal feed and vertical feed separately, their flatness is significantly lower than that of other surfaces.

(3) Parallel feed mode and reciprocating parallel feed mode (see Figure 4-2-9 (c) and Figure 4-2-9(d)). Parallel feed mode is to cut in one-way or reciprocating straight line in one direction. All the tool marks are straight parallel lines. The flatness accuracy by the one-way cutting is high, but its cutting efficiency is low (with empty tool stroke); the flatness accuracy by the reciprocating tool is low (due to the alternation in down milling and reverse milling), but the cutting efficiency is high. For large surfaces that require higher accuracy, one-way parallel feed mode is generally used.

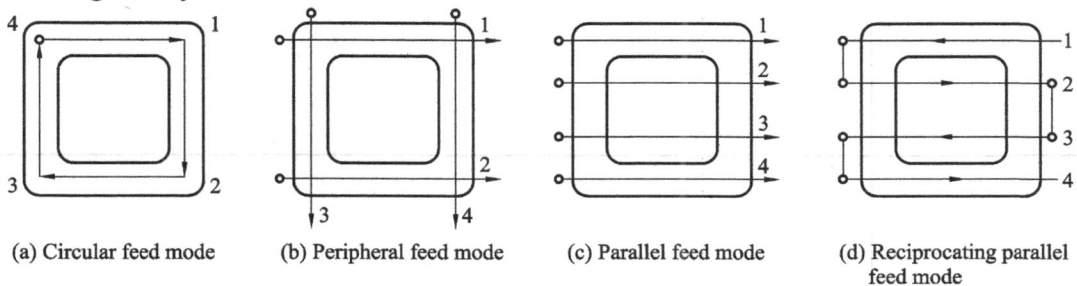

| (a) Circular feed mode | (b) Peripheral feed mode | (c) Parallel feed mode | (d) Reciprocating parallel feed mode |

Figure 4-2-9　Feeding Mode for Surface Milling

3. Precautions

The end milling cutter made of cemented and non-regrinding carbide should be selected in surface milling. Generally, a second run of cutting is adopted. It is better for the first run to be rough milling by end milling cutter, with continuous cutting along the surface of workpiece. Select the width of each run and the diameter of the milling cutter. The width of each run is recommended to be 60%-90% of the diameter of the tool, so that the joint mark does not affect the precision of the finish milling. If the machining allowance is large and uneven, the diameter of the milling cutter should be smaller. In finish milling, the diameter of the milling cutter should be larger, and it is a better operation that covering the entire width of the processed surface. In actual work, the semi-finish milling and finish milling are generally performed with indexable dense teeth face milling cutters or end milling cutters, which can achieve an ideal processing quality in surface milling, and can even replace grinding. The densely distributed cutter teeth greatly increase the feed speed, thereby improving the cutting efficiency. In finish milling, the end milling cutter can be set with 2-6 teeth.

4. Skills Training

Complete the task sheet as Table 4-2-1 based on what you have learned above.

Table 4-2-1 Task sheet

Task name	Automatic machining and dimensional accuracy control in surface milling of parts	
Class		Group No.
Task assignment		

Methods for checking the flatness of surfaces	
Methods for controlling dimensional accuracy	

Module 5

Basic Maintenance of Machining Center

Task 1　Maintenance Routine of Machining Center

【Knowledge objectives】

(1) Learn the methods for basic maintenance of machining center.

(2) Understand the knowledge of regular maintenance of machining center.

【Ability objectives】

(1) Master the methods for basic maintenance of machining center.

(2) Master the methods for regular maintenance of machining center.

1. Basic Maintenance of Machining Center

① Before each startup, check the input voltage of the machine tool, which should be $380\pm$ 10% V.

② The compressed air pressure must be 0.6 MPa. Check whether there is air leakage at any time.

③ Check the guide rail surface of X, Y, and Z axis at any time. If there are particles such as iron scraps attached to them, they should be removed in time. If there is any damage on the lead rail, it should be polished with oilstone.

④ Every time the tool is loaded on the machine tool, it must be checked in advance whether the pull pin is firmly installed on the tool handle.

⑤ Before starting up the machine, check the lubrication of the lead rail and ball screw. The lead rail and ball screw must be fully lubricated before running the machine. If the machine tool has not been running for a long time, press the automatic lubrication pump button several times to circulate the lubricant and ooze out of the guide rail and ball screw.

⑥ Having started up the machine tool, it has to first return to the reference point, and then runs at a low speed for 10–20 minutes. Check whether there is abnormal sound and vibration.

⑦ Each time the machine tool operates, it must be thoroughly cleaned, especially its guide rail operation panel. In addition, mechanical oil should be coated on the taper hole of the spindle and the taper handle of the cutter to prevent rusting, but the mechanical oil should be wiped off the machine before we restart the machine.

2. Regular Maintenance of Machining Center

① Check the oil level of the central lubricating station's oil tank every week, which should be higher than half. If the oil level is not as that high, the lubricating oil of a specified grade should be supplemented to reach 80% of the tank's capacity in time.

② Check the oil level of the spindle gear weekly, which should be kept half by the observation window.

③ Check the liquid level of the cooling tank weekly, which should be more than 3/4 of its capacity.

④ Clean the coolant filter screen once a month.

⑤ Check the oil scrapers on the lead rail surfaces of the X, Y, and Z axes every six months. Replace them immediately if damaged.

⑥ Change the coolant every six months.

⑦ Clean the filter screen of the centralized lubrication station every six months.

⑧ Adjust the inclined wedge of guide rail of X, Y and Z axes every six months.

⑨ Replace the gearbox oil every three years.

⑩ Replace the lubricating grease of the spindle bearing and axial bearing every three years.

3. Skills Training

Complete the task sheet as Table 5-1-1 based on what you have learned above.

Table 5-1-1 Task sheet

Task name	Routine maintenance of machining center		
Class		Group No.	
Task assignment			
Methods for basic maintenance			
Methods for regular maintenance			

Task 2   Maintenance System of Machining Center

【Knowledge objectives】

(1) Understand the basic maintenance system of machining center.

(2) Understand the secondary maintenance system of machining center.

【Ability objectives】

(1) Master the details of basic maintenance of machining center.

(2) Master the details of secondary maintenance machining center.

1. Basic Maintenance System of Machining Center

1) Mechanical maintenance

① Check lubricating system and pressure gauge, clean the filter screen of lubricating system, replace lubricating oil, and clear the oil passage.

② Check the gas circuit system, clean the air filter screen, and eliminate the leakage of pressure gas.

③ Check the liquid system, clean the filter, wash the oil tank, and replace or filter the oil. If possible, replace the seal.

④ Tighten all transmission parts and replace defective standard parts.

⑤ Apply lubricating grease to the grease lubrication part as required.

⑥ Clean and wash all transmission surfaces.

⑦ Check the tool bank and the manipulator, analyze the wear of the manipulator, and propose suggestions for replacement.

⑧ Check and repair damaged external components.

⑨ Check the protective cover.

2) Electrical maintenance

(1) Clean the electrical components in control cabinet, and check and fasten the terminals.

(2) Clean the control module, circuit board of CNC system, fan, air filter screen and heat sink.

(3) Clean the servo motor fan blades.

(4) Clean the internal components of operation panel, circuit board and fan. Check and fasten the connectors.

2. Secondary Maintenance System of Machining Center

1) Safety regulations

① The operator must carefully read and grasp the instruction signs such as danger, warning and caution on machine tool.

② When the protective cover, internal lock or other safety devices of the machine tool lose efficacy, the machine tool must be stopped.

③ The operator is strictly prohibited to modify the machine parameters.

④ In maintenance or processing other operations on the machine tool, it is strictly forbidden to lean your body under the workbench.

⑤ Before inspection, maintenance and repair, the power supply must be cut off.

⑥ It is strictly forbidden to overload, overrun, or operate the machine tool in violation of regulations.

⑦ In operating the CNC machine tool, the operator must be highly concentrated, and it is strictly forbidden to wear gloves and a tie, or to walk away without stopping the machine tool.

⑧ When there are workpieces, accessories or obstacles on the worktable, the magnifying power of rapid movement speed of each axis of the machine tool should be less than 50%.

2) Daily maintenance

Check the overall appearance of the equipment to find whether there are abnormal conditions, ensure that the equipment is clean and free of rust, and check whether the hydraulic system, pneumatic system, cooling device and power grid voltage are normal. After the machine started up, check whether the systems are normal, and run the spindle at a low speed for 5 min to observe whether the machining center is normal. Clean the taper hole of the spindle in time. Make sure that the site is clear after work finished.

3) Weekend maintenance

Fully clean the machine tool, inspect the appearance of cables and pipelines, and clean the outer surface structure of the spindle, the worktable, the surface of the manipulator, etc. Check whether the hydraulic pressure and cooling devices are normal, and clean the filter screen of the thermostatic device of spindle in time. Check the coolant, replace it if it is unqualified, and clean the device for removing chips.

3. Skills Training

Complete the task sheet as Table 5-2-1 based on what you have learned above.

Table 5-2-1   Task sheet

Task name	Maintenance system of machining center		
Class		Group No.	
Task assignment			
Primary maintenance system			
Secondary maintenance system			